Selected Works

Philip P Jackson

First Edition 2010

ISBN 978-1-4452-7964-0

Introduction

1. Algorithms

Introduction

Analysis

Results

2. General Topology

Transfinite Extensions

Ordinal Invariants

3. Forest Drainage

The Crychan Drainage Experiment After Felling

Hafren 4 Drainage Experiment

4. Database Development

Use of Rushmore Technology in Browse Edit Windows

A Custom Built General Reporting Facility

Introduction

This is a collection of works written between 1980 and 1998, whilst working on various projects after completing my PhD.

1. Algorithms for distribution-based number sorting was written as a dissertation for the degree of Master of Science. It analyses and tests the performance of an address calculation sorting algorithm and compares it with the standard number sorting algorithms.

2. Develops some topological ideas, which extend the groundwork laid in my thesis, "Iterated Remainders in Compactifications". In Transfinite Extensions an alternative proof is developed for certain theorems proved therein. Ordinal Invariants uses this theory and examines some ordinal quantifiers whose existence the theory implies.

3. Written for the Forestry Commission we develop a quantitative model of Forest Drainage and relate this to experimental observation. The analysis of the Crychan experiment introduces the techniques and prepares for the deeper analysis of the experiment at Hafren 4.

4. Written whilst working at the JET Joint Undertaking near Abingdon, this contains details of programs and programming techniques which can be used when developing database applications in the FoxPro language.

1. Algorithms

Introduction

Analysis

Testing

CONTENTS

1.1   Introduction

We are concerned with the problem of sorting an array of numbers into increasing (or decreasing) order by means of a computer algorithm. In particular we will investigate the case where the array of numbers is known to be generated as a random sample from some statistical distribution. Each such random sample, (of size N), will, in view of the ordering relation between the numbers, determine a permutation of size N. Also, if further random samples are taken then the distribution of the corresponding permutations is such that each permutation of length N will be equally likely to occur.

The sorting technique which we will use is known as address calculation sorting. Address calculation sorting was first studied in 1956 by Isaac and Singleton (3). The basic principle óf this technique is to use arithmetic operations to calculate approximately the position of each number in the sorted array. Then, by considering the neighbouring elements, the position is adjusted until eventually a fully sorted array emerges. These two stages of the process are sometimes refered to as coarse focusing and fine focusing.

Following Knuth (4) page 99 we will use linked linear lists to record the ordering of the elements in the array. This has two advantages, (i) it facilitate the keeping of several ordered lists of different lengths without any wastage of storage space, and (ii) it enables elements to be inserted into one of these ordered lists without shuffling along the existing elements of the list.

In chapter 2 we will determine analytically such properties of our algorithm as the expected number of comparison statements to be executed, and the variance of the number of comparison statements executed. To do this we need to consider our algorithm as a two stage process, and first investigate the case of a simple insertionsort algorithm using linked linear lists.

In chapter 3 we describe the development of a FORTRAN 77 program to implement the address calculation sorting algorithm. The average CPU time taken by the algorithm to sort samples of different sizes and from differing distributions is calculated. Concentrating on the uniform distribution, for which theoretically the smallest CPU times should be obtained, we will try to model the CPU times by a simple model based on the analytic investigation of chapter 2.

1.2 Linked linear lists

In a linked linear list the information about the ordering of a sequence s_1, s_2, .. , s_n is retained by associating with each s_i an integer valued pointer P_i that records the location at which s_{i+1} and P_{i+1} are stored. There is also a pointer P_0 which records the location of the head of the sequence, that is the location of s_1 and P_1.

Diagrammatically we have

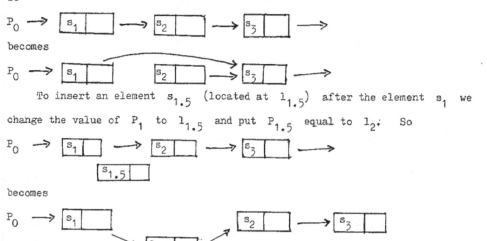

The use of this linked representation facilitates the operations of insertion of a sequence element after s_i and the deletion of a sequence element s_{i+1} if the location of s_i is known.

Thus to delete the element s_2 we simply change the value of P_1 to l_3. So

becomes

To insert an element $s_{1.5}$ (located at $l_{1.5}$) after the element s_1 we change the value of P_1 to $l_{1.5}$ and put $P_{1.5}$ equal to l_2. So

becomes

(See (5) page 39).

Also by using M different initial pointers we can partition an array of elements into M separate ordered lists.

1.3 Outline of the address calculation sorting algorithm.

An array X of N real numbers is to be sorted into ascending order by means of the algorithm.

The maximum, minimum, and hence range of X are first determined.

The interval $\begin{bmatrix} \min X, \max X \end{bmatrix}$ is considered to be partitioned into M equal length closed-open subintervals which are labelled 1 to M. For computational convenience a further 'interval' labelled M+1 is used which will contain only the maximum element(s) of X. The elements of X are thus partitioned into M+1 groups. If x is an element of X the label of the subinterval to which x belongs can be determined by the simple arithmetic formula $(((x - \min X)/\text{range } X)*M) + 1$ truncated to its integer part.

Each of the M+1 groups is sorted using an insertionsort algorithm. The ordering of the elements in each group is maintained by using linked linear lists. Thus for each element of X there is a corresponding integer valued pointer, and for each of the M+1 groups there is an integer valued initial pointer. For convenience an additional element X(0) is set to infinity and used to terminate each of the M+1 linked linear lists.

The elements of X are considered in turn. The address of the element X(I) is calculated, that is the label of the subinterval in which X(I) lies is determined. Let the address of X(I) be K. Then by comparing X(I) with the elements in list K so far considered the position of X(I) within the Kth list at this stage is determined (this is done as in an insertionsort algorithm).

This is done for each X(I), I=1,2,..,N until all the M+1 lists are complete. That is all the elements of X have been considered. Then the lists are used to construct an array W whose elements are those of X in ascending order.

Clearly the average CPU time taken by the algorithm will depend upon the distribution from which X is a random sample, and the value of M, the number of subintervals into which the range of X is partitioned. We will consider

the problem of choosing the best value of M and the effect of different

distributions in chapter 3.

CHAPTER 2

Before proceeding to an analytical investigation of the address calculation
sorting algorithm it is necessary to consider fully an insertionsort algorithm
which is used in sorting the M+1 lists generated by the address calculation
algorithm.

The principle of the insertionsort algorithm is as follows. Each element of
the array to be sorted is considered in turn. When the nth element of the
array is under consideration the previous n-1 elements have been put in
ascending order; this order being maintained by a linked linear list. The nth
element is then compared with the smallest element, then the second smallest
element, then the third smallest element, etc. until its position within the
first n ordered elements is determined. The nth element is then inserted
into this position, which is easily done because of the linked linear lists, and
the first n elements are now ordered. This process is continued until all the
elements of the array have been considered.

We give below a FORTRAN 77 subroutine which will sort an array Y of length
N using this insertionsort algorithm and produce an array W which contains the
elements of Y in ascending order. We set the element Y(0) to the value
infinity for computational convenience. This may be undesirable if it were to
be used as a practical working subroutine.

2.1 The Insertionsort Subroutine.

```
      SUBROUTINE  SORT(N,Y,P,W)
C     The array  Y(1:N)  to be sorted and the integer parameter  N  are obtained
C     from the main program.  The array  W(1:N) will after completion of the
C     subroutine return to the main program the elements of  Y(1:N)  in ascending
C     order as  W(1:N).   The integer array  P(0:N)  is used to store the pointers of
C     the linked linear lists.
      INTEGER  P(0:N), S, NEXT
      REAL  Y(0:N), W(0:N)
```

C   The location S is the initial pointer for the linked linear list. The
C pointers are first set to zero and Y(0) is set to the value infinity.

      Y(0) = 1E38
      DO 10, I=0, N
10   P(I) = 0
      S = 0
C   We begin sorting Y(1:N) by considering each element in turn.
      DO 50, J=1, N
C   If Y(J) is the smallest element so far considered then the initial pointer
C S is set to the value J and the Jth pointer P(J) is set to the previous
C value of S.
      IF(Y(J).LE.Y(S)) THEN   (1)
      P(J) = S
      S = J
      ELSE
C   If Y(J) is not the smallest element so far considered then Y(J) is
C compared with each (sorted) element in turn (i.e. with the second smallest,
C then with the third smallest, etc.) until an element larger than Y(J) is
C encountered. Then Y(J) is inserted into the linked linear list at the
C correct point.
      NEXT = S
30   IF(Y(J).LE.Y(P(NEXT))) GOTO 31   (2)
      NEXT = P(NEXT)
      GOTO 30
31   P(J) = P(NEXT)
      P(NEXT) = J
      END IF
50   CONTINUE
C   All the elements of Y(1:N) have now been considered and the linked linear
C list records them in ascending order. The array W(1:N) is set to contain
C the elements of Y in ascending order.

```
      NEXT = S
      DO 60, I=1, N
      W(I) = Y(NEXT)
 60   NEXT = P(NEXT)
      RETURN
      END
```

2.2 Analysis of the Insertionsort Subroutine.

We will analyse this insertionsort subroutine with a view to finding the expectation and variance of the number of times various comparison and assignment statements will be made when the array Y is assumed to be a random sample from some distribution.

First we require a definition and proposition from Combinatorial Theory which will be used in our analysis.

Definition Let $X = (x_1, x_2, \ldots, x_n)$ be a sequence of distinct real numbers. Then with respect to ascending order X may be considered as a permutation of size n. For each i, $1 \leq i \leq n$, put d_i = the number of x_j with $j < i$ such that $x_j > x_i$. That is d_i is the number of elements greater than x_i which occur to the left of x_i in the sequence. Then the vector $(d_1, d_2, \ldots, d_n)$ is called the inversion vector of the permutation X.

Proposition 1. Let S_n be the set of permutations of size n. Let $D = \left\{ (d_1, d_2, \ldots, d_n): d_i \in \mathbb{Z} \quad 0 \leq d_i < i \quad \text{for each } i \right\}$. Then there is a one-one correspondence between S_n and D in which each permutation is mapped to its inversion vector.

Proof See (5) pages 164-165.

We can now proceed with the analysis of the insertionsort subroutine.

Suppose the array Y to be sorted is $(y_1, y_2, \ldots, y_N)$ and that the y_i are distinct (as occurs with probability one when Y is a random sample from some continuous distribution). Let $(d_1, d_2, \ldots, d_N)$ be the inversion vector correspon to the permutation $(y_1, y_2, \ldots, y_N)$.

Consider the DO-LOOP, "DO 50, J=1, N". We consider which of its statements will be executed and how many times when J takes the value j.

Let the statement "IF(Y(J).LE.Y(S))THEN" be called comparison (1), and let
the statement "IF(Y(J).LE.Y(P(NEXT))) GOTO 31" be called comparison (2); and
the statements within the loop of the form "variable name = variable name"
be called assignment statements. Then considering the different possible values
of d_j we obtain the following table.

Let J = j.

Value of d_j	Number of executions of comparison (2)	Number of assignment statements executed
j-1	0	2
j-2	1	2+1
j-3	1+1	2+1+1
j-4	1+1+1	2+1+1+1
..	..	..
0	j-1	j-1+2

Also, clearly, comparison (1) will be executed once whatever the value of d_j.

Thus, when J = j,

(i)   comparison (1) will be executed once;

(ii)   comparison (2) will be executed $j-d_j-1$ times; and

(iii) $j-d_j+1$ assignment statements are executed.

Thus summing over J we find that to execute the program,

(i)   comparison (1) will be made a total of N times;

(ii)   comparison (2) will be made a total of

$$\sum_{j=1}^{N}(j-d_j-1) = \sum_{j=1}^{N} j - \sum_{j=1}^{N} d_j - N$$
$$= \tfrac{1}{2}N(N+1) - \sum_{j=1}^{N} d_j - N = \tfrac{1}{2}N(N-1) - \sum_{j=1}^{N} d_j \quad \text{times;}$$

and (iii) a total of $\sum_{j=1}^{N}(j-d_j+1) = \tfrac{1}{2}N(N+3) - \sum_{j=1}^{N} d_j$ assignment statements
will be executed.

Now for each j, $0 \leqslant d_j < j$, so in view of the proposition above
$$\max_{S_n}\left\{\sum_{j=1}^{N} d_j\right\} = \sum_{j=1}^{N}(j-1) = \tfrac{1}{2}N(N-1), \quad \text{and} \quad \min_{S_n}\left\{\sum_{j=1}^{N} d_j\right\} = 0.$$

So for the best case (when the array is already in ascending order), comparison
(1) is executed N times, comparison (2) is never executed, and $2N$ assignment
statements are executed.

For the worst case (when the array is in descending order), comparison (1)
is executed N times, comparison (2) is executed $\frac{1}{2}N(N-1)$ times, and $\frac{1}{2}N(N+3)$
assignment statements are executed.

Suppose now that we are sorting a random sample of size N from some
distribution. Then we know that each permutation in S_N is equally likely to
occur. Hence, in view of proposition 1, each inversion vector is equally likely
to occur. So d_j will take the values 0 to $j-1$ each with equal probability
and for $i \neq j$ d_i and d_j are independent.

Thus, in this situation, the expectation of $\sum\limits_{j=1}^{N} d_j$ will be

$$E\left(\sum\limits_{j=1}^{N} d_j\right) = \sum\limits_{j=1}^{N} E(d_j) = \sum\limits_{j=1}^{N} \frac{1}{2}(j-1)$$

(because d_j takes the values 0 to $j-1$ each with equal probability)

$$= \frac{1}{4}N(N-1).$$

Thus when sorting random samples of size N from some distribution we see
that, (i) comparison (1) is executed N times;

(ii) on average comparison (2) is executed $\frac{1}{4}N(N-1)$ times;

and (iii) on average $\frac{1}{2}N(N+3) - \frac{1}{4}N(N-1) = \frac{1}{4}N(N+7)$ assignment statements are
executed.

(c.f. (5) pages 281-283)

We also see that both the number of comparison statements executed and the
number of assignment statements executed will have the same variance. This is
the variance of $\sum\limits_{j=1}^{N} d_j$.

Now $\text{Var}\left(\sum\limits_{j=1}^{N} d_j\right) = \sum\limits_{j=1}^{N} \text{Var}(d_j)$ (because the d_j are independent)

Also $E(d_j^2) = \sum\limits_{x=0}^{j-1} x^2 \cdot \frac{1}{j} = \frac{1}{6}(j-1)(2j-1)$.

And $E(d_j) = \frac{1}{2}(j-1)$.

So $\text{Var}(d_j) = \frac{1}{6}(j-1)(2j-1) - \frac{1}{4}(j-1)^2$

$\qquad = \frac{1}{12}(j-1)(4j-2-3j+3) \qquad = \frac{1}{12}(j-1)(j+1)$

So $\text{Var}\left(\sum\limits_{j=1}^{N} d_j\right) = \frac{1}{12}\sum\limits_{j=1}^{N}(j-1)(j+1) = \frac{1}{12}\sum\limits_{j=1}^{N}(j^2 - 1)$

$\qquad = \frac{1}{12}\left(\frac{1}{6}N(N+1)(2N+1) - N\right)$

$$= \frac{1}{72}(2N^3 + 3N^2 + N - 6N) \qquad = \frac{1}{72}N(2N^2 + 3N - 5)$$

$$= \frac{1}{72}(N-1)N(2N+5)$$

This completes our analysis of the insertionsort subroutine.

2.3 The address calculation sorting algorithm.

We can now examine the address calculation algorithm in detail. The algorithm has already been outlined in chapter 1. We give here a FORTRAN 77   subroutine designed to implement the algorithm. The array to be sorted is X(1:N). As in the insertionsort subroutine we use X(0) set to infinity for convenience.

```
      SUBROUTINE ADSORT(N,X,W,M,P,LP,L)

      REAL  X(0:N),  W(0:N)

      INTEGER  P(0:N), LP(M+1), L(M+1), NEXT

C     The array  X  whose elements  X(1:N)  are to be sorted is obtained from the
C     main program together with its size  N.   The array  W  will, after execution
C     of the subroutine, return the elements of  X(1:N)  in ascending order as  W(1:N).
C     The value of  M  will be the number of subintervals into which the range of
C     X  will be partitioned.  The pointers for the linked linear lists are stored in
C     P(0:N); and the initial pointers for the  M+1  lists to be constructed are
C     contained in LP.  The integer array  L  is used to record the number of
C     elements in each of the  M+1  lists.
C     We first find the maximum, minimum, and range of the array  X.

      XMAX = X(1)

      XMIN = X(1)

      DO 10, I=2, N

      IF(X(I).GT.XMAX)THEN

      XMAX = X(I)

      ELSE

      IF(X(I).LT.XMIN) XMIN = X(I)

      END IF

  10  CONTINUE

      RANGE = XMAX - XMIN
```

```
C    The pointers  P, initial pointers  LP  and the integer array  L  are set to
C  zero;  X(O)  is set to the value infinity.
     DO 18, I=0, N
 18  P(I) = 0
     DO 19, I=1, M+1
     L(I) = 0
 19  LP(I) = 0
     X(0) = 1E38
C    We now begin to construct the  M+1  lists.
     DO 50, J=1, N
C    The address  K  of the list into which  X(J)  is to be inserted is first
C  calculated.
     K = ((X(J) - XMIN)/RNGE)*M
     K = K + 1
C    An accumulating tally of the number of elements in each list is kept.
     L(K) = L(K) + 1
C    We now insert  X(J)  into list  K  in the manner of the insertionsort
C  algorithm.
C    Is  X(J)  the smallest element so far in list  K?
     IF(X(J).LE.X(LP(K))) THEN
     P(J) = LP(K)
     LP(K) = J
     ELSE
C    If  X(J)  is not the smallest so far in list  K,  we compare  X(J)  with
C  each element of list  K  in turn until an element greater than  X(J)  is
C  encountered.
     NEXT = LP(K)
 30  IF(X(J).LE.X(P(NEXT))) GOTO 31
     NEXT = P(NEXT)
     GOTO 30
 31  P(J) = P(NEXT)
```

```
        P(NEXT) = J

        END IF

50   CONTINUE

C    The M+1 lists have now been constructed. We use these lists to assign

C    the elements of  X(1:N)  to the array  W(1:N)  so that  W  is in ascending

C    order.

        I = 1

        DO 60, K=1, M+1

        NEXT = LP(K)

        DO 60, J=1, L(K)

        W(I) = X(NEXT)

        NEXT = P(NEXT)

60   I = I + 1

        RETURN

        END
```

2.4 Analysis of the address calculation sorting algorithm.

2.4.1   We will now derive the expected number of times comparison statements
and assignment statements will be executed when running this subroutine with a
random sample from some distribution, as we did for the insertionsort algorithm.

Suppose that $\underline{X}$ is a random sample of size N from a known distribution.
Let the range of $\underline{X}$ be "partitioned" into m closed-open intervals $K_1, \ldots, K_m$.
Let $\underline{X}_i$ denote the vector whose components are those elements of $\underline{X}$ which lie
in the interval K_i (with the ordering preserved as for $\underline{X}$). Let n_i be the
length of $\underline{X}_i$ for each i. (We ignore here the fact that $\max \underline{X} \notin K_i$ for any
i since the effect of this is negligible.) So $\sum_{i=1}^{m} n_i = N$.

Let $\underline{n} = (n_1, n_2, \ldots, n_m)$. Then since we have a random sample from a fixed
distribution, ignoring the (negligible) fact that $\min \underline{X} \in K_1$ always, we may
assume that $\underline{n}$ will have a multinomial distribution $M(N, \underline{p})$ where
$\underline{p} = (p_1, p_2, \ldots, p_m)$, $\sum_{i=1}^{m} p_i = 1$, which depends upon the distribution from which

the sample is taken.

Also for fixed i, given n_i, the permutation of length n_i which corres[ponds]
to the vector $\underline{X}_i$ will be such that all permutations of length n_i will be equa[lly]
likely to occur.

Now for each i let C_i denote the number of comparison statements within
the DO-LOOP, "DO 50, J=1, N" which will be executed whilst sorting list i;
that is whilst sorting the vector $\underline{X}_i$. Then from the analysis of the insertion[sort]
subroutine we know that for fixed i, $E(C_i \mid n_i) = n_i + \frac{1}{4}n_i(n_i - 1)$.

Hence
$$E(C_i) = E(E(C_i \mid n_i))$$
$$= E(n_i) + \frac{1}{4}E(n_i(n_i - 1))$$
$$= Np_i + \frac{1}{4}N(N - 1)p_i^2$$

Hence the expected number of times comparison statements are executed within
the DO-LOOP, "DO 50, J=1, N" to sort the whole of the array X will be

$$E\left(\sum_{i=1}^{m}C_i\right) = \sum_{i=1}^{m}E(C_i) = N\sum_{i=1}^{m}p_i + \frac{1}{4}N(N - 1)\sum_{i=1}^{m}p_i^2$$
$$= \frac{1}{4}N(N - 1)\sum_{i=1}^{m}p_i^2 + N \qquad (*)$$

Similarly, let A_i denote the number of assignment statements executed
within the DO-LOOP, "DO 50, J=1, N", (after the statement $L(K) = L(K) + 1$)
to sort list i, that is $\underline{X}_i$. Then from the analysis of the insertionsort
subroutine we know that $E(A_i \mid n_i) = \frac{1}{4}n_i(n_i - 1) + 2n_i$.

So $E(A_i) = E(E(A_i \mid n_i))$
$$= \frac{1}{4}E(n_i(n_i - 1)) + 2E(n_i) = \frac{1}{4}N(N - 1)p_i^2 + 2Np_i$$

Hence the expected number of times these assignment statements are executed
whilst sorting the whole of X will be

$$E\left(\sum_{i=1}^{m}A_i\right) = \frac{1}{4}N(N - 1)\sum_{i=1}^{m}p_i^2 + 2N \qquad (**).$$

2.4.2 Special case when X is uniformly distributed.

Now from (*) and (**) we see that the address calculation subroutine will be
most efficient when X is a random sample from that distribution for which
$\sum_{i=1}^{m}p_i^2$ is minimised. From proposition 2 below we see that this is so when
$p_i = 1/m$ for each i. Thus the address calculation subroutine is theoretically
most efficient when X is a random sample from the uniform distribution.

Proposition 2 Let $\sum_{i=1}^{m} p_i = 1$, $0 \leqslant p_i \leqslant 1$ for all i. Then $\sum_{i=1}^{m} p_i^2$ is minimised when $p_i = 1/m$ for each i.

Proof Let λ be a Lagrange multiplier. We must minimise

$$\Omega = \sum_{i=1}^{m} p_i^2 - \lambda \left(\sum_{i=1}^{m} p_i - 1 \right).$$

So setting the derivative with respect to p_i to zero we have $\lambda = 2p_i$.

Hence $\sum_{i=1}^{m} \lambda = 2 \sum_{i=1}^{m} p_i$

i.e. $m\lambda = 2$

i.e. $\lambda = 2/m$

So $p_i = \lambda/2 = 1/m$ for a minimum.

When X is a sample from the uniform distribution the formulae (*) and (**) simplify to the following:

E(# of comparison statements executed) $= \dfrac{N(N - 1)}{4M} + N$, and

E(# of assignment statements executed) $= \dfrac{N(N - 1)}{4M} + 2N$.

2.4.3 Analysis of the range calculation

We also examine that part of the program in which the maximum, minimum, and range of X are determined. Clearly the comparison "IF(X(I).GT.XMAX)THEN" will be made N-1 times. Also let (d_i) be the inversion vector corresponding to X. Then, for each i, X(I) is greater than XMAX if and only if $d_i = 0$. So the comparison "IF(X(I).LT.XMIN)" will be made whenever $d_i \neq 0$, and the expected number of times this comparison is executed will be

$\sum_{i=2}^{N} \text{Prob}\{d_i \neq 0\} = \sum_{i=2}^{N}(i-1)/i = N - 1 - \sum_{i=2}^{N} 1/i$

But $\sum_{i=1}^{N} 1/i \leqslant \int_{1}^{N} \dfrac{dx}{x} = \log N$

Hence, for large N, $\sum_{i=2}^{N} 1/i$ is negligible when compared with N.

e.g. when N = 2000, $\sum_{i=2}^{N} 1/i \leqslant 7.6$.

So the comparison "IF(X(I).LT.XMIN)" will be made on average approximately N-1 times for large N.

2.4.4 Variance of the number of comparison statements executed when X is uniformly distributed.

We have seen that when X is a random sample from the uniform distribution $E(\sum_{i=1}^{m} C_i) = \dfrac{N(N-1)}{4M} + N$. We will now derive a formula for $Var(\sum_{i=1}^{m} C_i)$ when X is a random sample from the uniform distribution.

We require the following results:

Let $Z \sim B(N, p)$. Let μ_r' denote the rth moment of Z about the origin, i.e. $\mu_r' = E(Z^r)$; and let $\mu_{(r)}'$ denote the rth factorial moment of Z, i.e. $\mu_{(r)}' = E(Z(Z-1)\ldots(Z-r+1))$.

The following are easily verified.

$\mu_1' = \mu_{(1)}'$

$\mu_2' = \mu_{(2)}' + \mu_{(1)}'$

$\mu_3' = \mu_{(3)}' + 3\mu_{(2)}' + \mu_{(1)}'$

$\mu_4' = \mu_{(4)}' + 6\mu_{(3)}' + 7\mu_{(2)}' + \mu_{(1)}'$

$\mu_{(1)}' = Np$

$\mu_{(2)}' = N(N-1)p^2 = (N^2-N)p^2$

$\mu_{(3)}' = N(N-1)(N-2)p^3 = (N^3-3N^2+2N)p^3$

$\mu_{(4)}' = N(N-1)(N-2)(N-3)p^4 = (N^4-6N^3+11N^2-6N)p^4$

Let, as before, C_i denote the number of comparison statements executed whilst sorting list i.

Then, by a well known result,

$$Var(\sum_{i=1}^{m} C_i) = Var(E(\sum_{i=1}^{m} C_i | \underline{n})) + E(Var(\sum_{i=1}^{m} C_i | \underline{n}))$$
$$= Var(\sum_{i=1}^{m} E(C_i | \underline{n})) + \sum_{i=1}^{m} E(Var(C_i | \underline{n}))$$

(because given $\underline{n}$, C_i and C_j are independent).

We have already seen that

$$E(C_i | \underline{n}) = \tfrac{1}{4}n_i(n_i+3) = \tfrac{1}{4}n_i^2 + \tfrac{3}{4}n_i$$

So $\sum_{i=1}^{m} E(C_i | \underline{n}) = \tfrac{1}{4}\sum_{i=1}^{m} n_i^2 + \tfrac{3}{4}\sum_{i=1}^{m} n_i = \tfrac{1}{4}\sum_{i=1}^{m} n_i^2 + \tfrac{3}{4}N$.

So $Var(\sum_{i=1}^{m} E(C_i | \underline{n})) = \dfrac{1}{16}Var(\sum_{i=1}^{m} n_i^2)$

Also, we have already seen that

$$Var(C_i | \underline{n}) = n_i(n_i-1)(2n_i+5)/72 = \dfrac{1}{72}\left\{ 2n_i^3 + 3n_i^2 - 5n_i \right\}$$

So $\sum_{i=1}^{m} E(\mathrm{Var}(C_i|\underline{n})) = \frac{1}{72}\sum_{i=1}^{m}\left\{2E(n_i^3) + 3E(n_i^2) - 5E(n_i)\right\}$

Since X is a random sample from the uniform distribution we have that $n_i \sim B(N, 1/m)$.

So $E(n_i) = N/m$

$E(n_i^2) = \mu_2' = \mu_{(2)}' + \mu_{(1)}' = (N^2-N)/m^2 + N/m$

$E(n_i^3) = \mu_3' = \mu_{(3)}' + 3\mu_{(2)}' + \mu_{(1)}'$

$\qquad = (N^3-3N^2+2N)/m^3 + 3(N^2-N)/m^2 + N/m$

So $\sum_{i=1}^{m} E(\mathrm{Var}(C_i|\underline{n})) = 2(N^3-3N^2+2N)/72m^2 + 6(N^2-N)/72m + 2N/72$

$\qquad\qquad + 3(N^2-N)/72m + 3N/72 - 5N/72$

$\qquad\qquad = 2(N^3-3N^2+2N)/72m^2 + 9(N^2-N)/72m$

Now $\mathrm{Var}\left(\sum_{i=1}^{m} n_i^2\right) = \sum_{i=1}^{m}\mathrm{Var}(n_i^2) + \sum_{i\neq j}\mathrm{Cov}(n_i^2, n_j^2)$

$\mathrm{Var}(n_i^2) = E(n_i^4) - (E(n_i^2))^2$

$\qquad = \mu_4' - (\mu_2')^2$

$\qquad = \mu_{(4)}' + 6\mu_{(3)}' + 7\mu_{(2)}' + \mu_{(1)}' - (\mu_{(2)}' + \mu_{(1)}')^2$

$\qquad = (N^4-6N^3+11N^2-6N)/m^4 + 6(N^3-3N^2+2N)/m^3 + 7(N^2-N)/m^2 + N/m$

$\qquad\quad - (N^4-2N^3+N^2)/m^4 - N^2/m^2 - 2N(N^2-N)/m^3$

$\qquad = (-4N^3+10N^2-6N)/m^4 + (4N^3-16N^2+12N)/m^3 + (6N^2-7N)/m^2 + N/m$

So $\sum_{i=1}^{m}\mathrm{Var}(n_i^2) = (-4N^3+10N^2-6N)/m^3 + (4N^3-16N^2+12N)/m^2 + (6N^2-7N)/m + N$

Also $\mathrm{Cov}(n_i^2, n_j^2) = E(n_i^2 n_j^2) - E(n_i^2)E(n_j^2)$

$E(n_i(n_i-1)n_j(n_j-1)) = E(n_i^2 n_j^2) - E(n_i n_j^2) - E(n_i^2 n_j) + E(n_i n_j)$

$E(n_i(n_i-1)n_j) = E(n_i^2 n_j) - E(n_i n_j)$

$E(n_i n_j(n_j-1)) = E(n_i n_j^2) - E(n_i n_j)$

It is easy to see that

$E(n_i(n_i-1)n_j(n_j-1)) = N(N-1)(N-2)(N-3)/m^4 = (N^4-6N^3+11N^2-6N)/m^4$

Similarly $E(n_i(n_i-1)n_j) = E(n_i n_j(n_j-1)) = (N^3-3N^2+2N)/m^3$

and $E(n_i n_j) = (N^2-N)/m^2$

So $E(n_i^2 n_j^2) = E(n_i(n_i-1)n_j(n_j-1)) + E(n_i n_j^2) + E(n_i^2 n_j) - E(n_i n_j)$

$= E(n_i(n_i-1)n_j(n_j-1)) + E(n_i n_j(n_j-1)) + E(n_i n_j) + E(n_i(n_i-1)n_j)$

$\quad + E(n_i n_j) - E(n_i n_j)$

$= (N^4-6N^3+11N^2-6N)/m^4 + 2(N^3-3N^2+2N)/m^3 + (N^2-N)/m^2$

and $E(n_i^2)E(n_j^2) = ((N^2-N)/m^2 + N/m)^2$

$= (N^4-2N^3+N^2)/m^4 + 2(N^3-N^2)/m^3 + N^2/m^2$

So $\mathrm{Cov}(n_i^2, n_j^2) = (-4N^3+10N^2-6N)/m^4 + (-4N^2+4N)/m^3 - N/m^2$

So $\sum_{i \neq j} \mathrm{Cov}(n_i^2, n_j^2) = m(m-1)\mathrm{Cov}(n_i^2, n_j^2)$ (for some $i \neq j$)

$= (m-1)\left\{(-4N^3+10N^2-6N)/m^3 + (-4N^2+4N)/m^2 - N/m\right\}$

$= (4N^3-10N^2+6N)/m^3 + (-4N^3+10N^2-6N+4N^2-4N)/m^2$

$\quad + (-4N^2+4N+N)/m - N$

$= (4N^3-10N^2+6N)/m^3 + (-4N^3+14N^2-10N)/m^2 + (-4N^2+5N)/m -$

So $\mathrm{Var}(\sum_{i=1}^{m} n_i^2) = \sum_{i=1}^{m}\mathrm{Var}(n_i^2) + \sum_{i \neq j}\mathrm{Cov}(n_i^2, n_j^2)$

$= (-4N^3+10N^2-6N+4N^3-10N^2+6N)/m^3 + (4N^3-16N^2+12N-4N^3+14N^2-10N)/m$

$\quad + (6N^2-7N-4N^2+5N)/m + N - N$

$= (-2N^2+2N)/m^2 + (2N^2-2N)/m$

So $\mathrm{Var}(\sum_{i=1}^{m} E(C_i|\underline{n})) = \frac{1}{16}\mathrm{Var}(\sum_{i=1}^{m} n_i^2) = (-N^2+N)/8m^2 + (N^2-N)/8m$

So $\mathrm{Var}(\sum_{i=1}^{m} C_i) = (-9N^2+9N+2N^3-6N^2+4N)/72m^2 + (9N^2-9N+9N^2-9N)/72m$

$= \frac{1}{72}\left\{(2N^3-15N^2+13N)/m^2 + (18N^2-18N)/m\right\}$

Thus when X is a random sample from the uniform distribution of size N the variance of the number of comparison statements executed is

$$\frac{1}{72}\left\{(2N^3-15N^2+13N)/M^2 + (18N^2-18N)/M\right\}.$$

3.1 Testing the address calculation sorting algorithm.

In this chapter we will examine the results obtained in practice when using the address calculation algorithm with various values of the parameter M and various distributions. To do this we have calculated the average CPU time taken for the algorithm to execute by means of the FORTRAN 77 program SORTTIME.F77, which is listed in the appendix. We use NAG library routines to generate samples from the various distributions; and the Prime Applications Library clock to evaluate the CPU time elapsed.

3.2 Uniformly distributed samples.

We put $M = \alpha N$ and consider which values of α will lead to minimising the elapsed CPU time, where N is the sample size and M is the number of intervals into which the range of X is "partitioned".

We have plotted the average CPU times obtained when SORTTIME is compiled with the Prime compiler on page 20.

We see that the minimum CPU times are obtained when α lies between 0.3 and 0.6. The elapsed CPU time rises steeply when α is less than 0.2.

In chapter 2 we have seen that when samples are taken from the uniform distribution

$$E(\# \text{ of comparison statements executed}) = N(N-1)/4M + N$$
$$= N(N-1)/4\alpha N + N$$
$$\simeq N/4\alpha + N \ ,$$

and similarly

$$E(\# \text{ of assignment statements executed}) = N(N-1)/4M + 2N$$
$$= N(N-1)/4\alpha N + 2N$$
$$N/4\alpha + 2N \ .$$

Also when finding the range of X the expected number of comparison statements executed is approximately 2N; and it is clear that the operations of setting the pointers to zero, calculating K and L(K) and constructing the vector W require executing statements of the same type approximately N or M $(=\alpha N)$ times on average.

This suggests we may be able to model the average CPU time elapsed by a model of the form $T(N, \alpha) = (A/\alpha + B\alpha + C)N$, where A, B, and C are constants.

To fit this model we have plotted the average CPU time taken when $\alpha = 0.1$ $\alpha = 0.5$, and $\alpha = 1.0$, for $N = 1000$, $N = 2000$, $N = 3000$, $N = 4000$, and $N = 5000$. Under the assumptions of the model, for fixed α, these points should lie on a straight line through the origin. We have fitted by the method of least squares the best straight lines through the origin and see that they provide a good fit.

(Note: Given s points (x_i, y_i) $1 \leq i \leq s$, to fit by least squares the line $y = mx$ we must estimate m so that $\sum_{i=1}^{s}(y_i - mx_i)^2$ is minimised. To do this it is easy to see that we must take $\hat{m} = \sum_{i=1}^{s}x_i y_i / \sum_{i=1}^{s}x_i^2$.)

Equating the gradients of these least squares fitted lines to $A/\alpha + B\alpha + C$ we obtain

$$10A + 0.1B + C = 5.0118e$$
$$2A + 0.5B + C = 4.018e$$
$$A + B + C = 4.2396e \qquad \text{where} \quad e = 10^{-4}$$

$\Longleftrightarrow$
$$100A + B + 10C = 50.118e$$
$$4A + B + 2C = 8.036e$$
$$A + B + C = 4.2396e$$

$\Longleftrightarrow$
$$90A - 9B = 7.722e$$
$$2A - B = -0.4432e$$
$$A + B + C = 4.2396$$

$\Longleftrightarrow$
$$72A = 11.7108e$$
$$2A - B = -0.4432e$$
$$A + B + C = 4.2396e$$

$\Longleftrightarrow$
$$\hat{A} = 0.16265e$$
$$\hat{B} = 0.7685e$$
$$\hat{C} = 3.30845e$$

We have plotted the fitted curves $T(N, \alpha) = (\hat{A}/\alpha + \hat{B}\alpha + \hat{C})N$ on page 24. We see that these are a good fit.

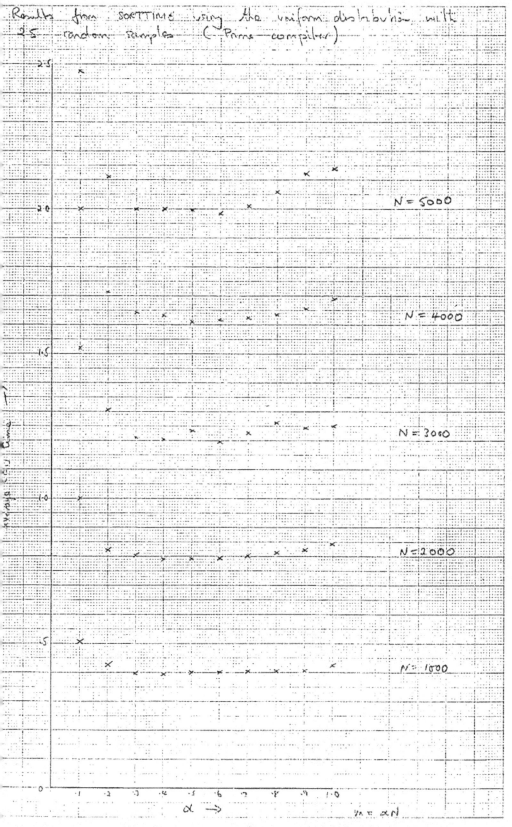

Results from SORTTIME using the uniform distribution with 25 random samples (Prime compiler)

$N = 5000$

$N = 4000$

$N = 3000$

$N = 2000$

$N = 1000$

$\alpha \longrightarrow$

$y_n = \alpha n$

RESULTS OF TEST
SIZE OF SAMPLES: N = 1000
USING THE UNIFORM   DISTRIBUTION
WITH PARAMETERS A = 0.00   , B =   1.0
AVERAGE CPU TIME TAKEN OVER   25 RANDOM SAMPLES

VALUE OF M	AVERAGE CPU TIME
100	0.5084
200	0.4258
300	0.3984
400	0.3948
500	0.3995
600	0.4000
700	0.4008
800	0.4048
900	0.4080
1000	0.4212

RESULTS OF TEST
SIZE OF SAMPLES: N = 2000
USING THE UNIFORM   DISTRIBUTION
WITH PARAMETERS A = 0.00   , B =   1.0
AVERAGE CPU TIME TAKEN OVER   25 RANDOM SAMPLES

VALUE OF M	AVERAGE CPU TIME
200	1.0080
400	0.8396
600	0.8024
800	0.7892
1000	0.7892
1200	0.7920
1400	0.7976
1600	0.8132
1800	0.8236
2000	0.8420

RESULTS OF TEST
SIZE OF SAMPLES: N = 3000
USING THE UNIFORM   DISTRIBUTION
WITH PARAMETERS A = 0.00   , B =   1.0
AVERAGE CPU TIME TAKEN OVER   25 RANDOM SAMPLES

VALUE OF M	AVERAGE CPU TIME
300	1.5484
600	1.3064
900	1.2124
1200	1.2052
1500	1.2340
1800	1.1984
2100	1.2384
2400	1.2612
2700	1.2424
3000	1.2488

```
                    RESULTS OF TEST
SIZE OF SAMPLES: N =  4000
USING THE   UNIFORM    DISTRIBUTION
WITH PARAMETERS  A =  0.00    , B =   1.0
AVERAGE CPU TIME TAKEN OVER    25 RANDOM SAMPLES

VALUE OF M              AVERAGE CPU TIME

    400                     2.0008
    800                   · 1.7132
   1200                     1.6428
   1600                     1.6308
   2000                     1.6100
   2400                     1.6172
   2800                     1.6244
   3200                     1.6344
   3600                     1.6532
   4000                     1.6900


                    RESULTS OF TEST
 SIZE OF SAMPLES: N =  5000
 USING THE   UNIFORM    DISTRIBUTION
 WITH PARAMETERS  A =  0.00    , B =   1.0
 AVERAGE CPU TIME TAKEN OVER    25 RANDOM SAMPLES

 VALUE OF M              AVERAGE CPU TIME

    500                     2.4784
   1000                     2.1116
   1500                     1.9973
   2000                     2.0055
   2500                     1.9955
   3000                     1.9816
   3500                     2.0132
   4000                     2.0636
   4500                     2.1240
   5000                     2.1412
```

Plot of average CPU time taken using uniform distribution over 25 random samples with $M = N$, $M = \frac{1}{2}N$ and $M = \frac{1}{10}N$ together with best 'least squares' straight line through the origin. (Prime compiler).

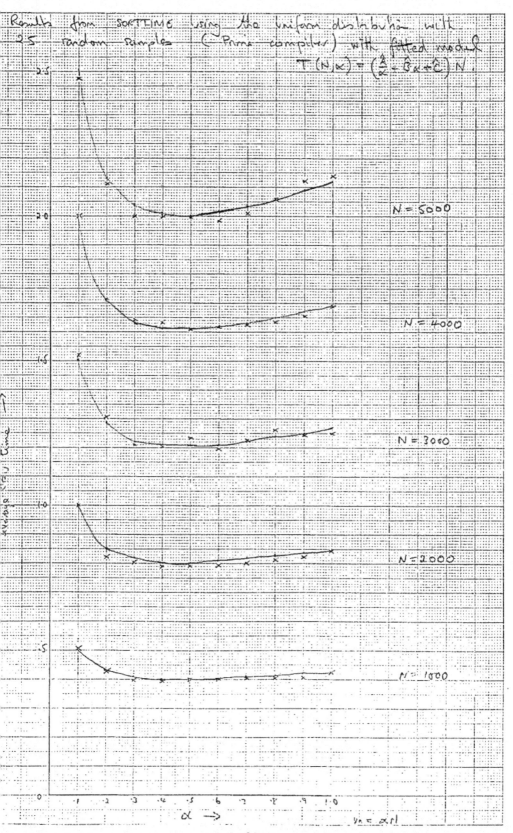

Results from SORTTIME using the uniform distribution with 25 random samples ('Pura' compiler) with fitted model

$$T(N, \alpha) = \left(\frac{\hat{A}}{\alpha} + \hat{B}\alpha + \hat{C}\right) N.$$

average CPU time →

2·5

2·0   N = 5000

N = 4000

·5

N = 3000

1·0

N = 2000

·5

N = 1000

0

·1   ·2   ·3   ·4   ·5   ·6   ·7   ·8   ·9   1·0

$\alpha \rightarrow$ $y_n = \alpha r_n$

— 24 —

We can minimise $T(N, \alpha)$ with respect to α for fixed N by putting

$\frac{dT}{d\alpha} = 0.$

i.e. $-A/\alpha^2 + B = 0$

i.e. $\alpha^2 = A/B$

i.e. $\alpha = \sqrt{A/B}$

We have $\hat{A} = 0.16265e$ and $\hat{B} = 0.7685e.$ So for a minimum $\hat{\alpha} = \sqrt{\hat{A}/\hat{B}} = 0.4$

Salford Compiler

We have also executed the program SORTTIME with samples from the uniform distribution when SORTTIME is compiled by means of the Salford compiler. These results are plotted on pages 27 and 28.

We see, as before, that to minimise the average CPU time elapsed we require to choose α between 0.3 and 0.6, and that the time rises steeply for α less than 0.2. However, due to better code generation, the times themselves are much quicker, e.g. when $N = 1000$ the minimum CPU time is approximately 0.13 seconds compared with 0.40 seconds when the prime compiler is used.

We have tried fitting the model $T(N, \alpha) = (A/\alpha + B\alpha + C)N$ by the same procedure to these new results. We notice that the least squares straight lines through the origin are not as good a fit as before. However using them we can estimate the parameters thus:

$10A + 0.1B + C = 1.624e$

$2A + 0.5B + C = 1.3889e$

$A + B + C = 1.4546e$

$\Longrightarrow \quad 100A + B + 10C = 16.24e$

$4A + B + 2C = 2.7778e$

$A + B + C = 1.4546e$

$\Longrightarrow \quad 90A - 9B = 1.694e$

$2A - B = -0.1314e$

$A + B + C = 1.4546e$

$$72A = 2.8766e$$

$$2A - B = -0.1314e$$

$$A + B + C = 1.4546e$$

$\Longleftrightarrow$
$$\hat{A} = 0.03995e$$

$$\hat{B} = 0.21131e$$

$$\hat{C} = 1.20334e$$

We have superimposed the fitted model onto our results. We see that the fit is O.K. for $N > 2000$ but for $N \leqslant 2000$ the model does not give so good a fit.

We have also plotted the variances for the CPU times obtained from SORTTIME using the Salford compiler and the uniform distribution. This plot on page 35 shows no clear relationship between the variance and the sample size N. All we can say is that relatively large variances seem to occur when $N > 3000$.

The value of α which minimises $T(N, \alpha)$ in this case is given by
$$\hat{\alpha} = \sqrt{\hat{A}/\hat{B}} = 0.43.$$

Results from soartime - using the uniform distribution
with 10 random samples   (Salford compiler)

N = 2000

N = 1500

N = 1000

N = 500

averages in time →

α →

M = α N

Results from SORTTIME using the uniform distribution
with 10 random samples (Salford compiler)

$N = 4000$

$N = 3500$

$N = 3000$

$N = 2500$

Time (seconds) →

$\alpha \rightarrow$ $M = \alpha N$

```
                    RESULTS OF TEST
SIZE OF SAMPLES: N =   500
USING THE   UNIFORM    DISTRIBUTION
WITH PARAMETERS   A =   0.00D+00 B =   1.0
AVERAGE CPU TIME TAKEN OVER    10 RANDOM SAMPLES
```

VALUE OF M	AVERAGE CPU TIME	VARIANCE
50	0.0720	0.000017778
100	0.0620	0.000017778
150	0.0610	0.000032222
200	0.0650	0.000050000
250	0.0630	0.000023333
300	0.0680	0.000017778
350	0.0690	0.000032222
400	0.0680	0.000017778
450	0.0670	0.000023333
500	0.0710	0.000010000

```
                    RESULTS OF TEST
SIZE OF SAMPLES: N =   1000
USING THE   UNIFORM    DISTRIBUTION
WITH PARAMETERS   A =   0.00D+00 B =   1.0
AVERAGE CPU TIME TAKEN OVER    10 RANDOM SAMPLES
```

VALUE OF M	AVERAGE CPU TIME	VARIANCE
100	0.1470	0.000023333
200	0.1280	0.000017778
300	0.1290	0.000032222
400	0.1280	0.000017778
500	0.1300	0.000000000
600	0.1330	0.000023333
700	0.1350	0.000027778
800	0.1370	0.000045556
900	0.1410	0.000010000
1000	0.1460	0.000115556

```
                    RESULTS OF TEST
SIZE OF SAMPLES: N =   1500
USING THE   UNIFORM    DISTRIBUTION
WITH PARAMETERS   A =   0.00D+00 B =   1.0
AVERAGE CPU TIME TAKEN OVER    10 RANDOM SAMPLES
```

VALUE OF M	AVERAGE CPU TIME	VARIANCE
150	0.2250	0.000050000
300	0.2010	0.000010000
450	0.1980	0.000040000
600	0.1940	0.000026667
750	0.1990	0.000032222
900	0.1990	0.000010000
1050	0.2060	0.000026667
1200	0.2100	0.000044444
1350	0.2090	0.000010000
1500	0.2130	0.000023333

```
                    RESULTS OF TEST
SIZE OF SAMPLES: N =  2000
USING THE   UNIFORM     DISTRIBUTION
WITH PARAMETERS   A =   0.00D+00 B =   1.0
AVERAGE CPU TIME TAKEN OVER    10 RANDOM SAMPLES
```

VALUE OF M	AVERAGE CPU TIME	VARIANCE
200	0.3310	0.000032222
400	0.2980	0.000040000
600	0.2870	0.000023333
800	0.2850	0.000027778
1000	0.2850	0.000027778
1200	0.2910	0.000076667
1400	0.2900	0.000066667
1600	0.2980	0.000017778
1800	0.3000	0.000022222
2000	0.3030	0.000023333

```
                    RESULTS OF TEST
SIZE OF SAMPLES: N =  2500
USING THE   UNIFORM     DISTRIBUTION
WITH PARAMETERS   A =   0.00D+00 B =   1.0
AVERAGE CPU TIME TAKEN OVER    10 RANDOM SAMPLES
```

VALUE OF M	AVERAGE CPU TIME	VARIANCE
250	0.4040	0.000026667
500	0.3620	0.000040000
750	0.3480	0.000040000
1000	0.3520	0.000062222
1250	0.3520	0.000017778
1500	0.3520	0.000040000
1750	0.3560	0.000048889
2000	0.3620	0.000062222
2250	0.3730	0.000045556
2500	0.3700	0.000044444

```
                    RESULTS OF TEST
SIZE OF SAMPLES: N =  3000
USING THE   UNIFORM     DISTRIBUTION
WITH PARAMETERS   A =   0.00D+00 B =   1.0
AVERAGE CPU TIME TAKEN OVER    10 RANDOM SAMPLES
```

VALUE OF M	AVERAGE CPU TIME	VARIANCE
300	0.4870	0.000045556
600	0.4270	0.000023333
900	0.4150	0.000050000
1200	0.4230	0.000134444
1500	0.4190	0.000098889
1800	0.4180	0.000040000
2100	0.4220	0.000040000
2400	0.4250	0.000050000
2700	0.4360	0.000160000
3000	0.4360	0.000026667

RESULTS OF TEST
SIZE OF SAMPLES: N = 3500
USING THE UNIFORM DISTRIBUTION
WITH PARAMETERS A = 0.00D+00 B = 1.0
AVERAGE CPU TIME TAKEN OVER 10 RANDOM SAMPLES

VALUE OF M	AVERAGE CPU TIME	VARIANCE
350	0.5620	0.000040000
700	0.5010	0.000232222
1050	0.4840	0.000048889
1400	0.4830	0.000045556
1750	0.4790	0.000054444
2100	0.4830	0.000023333
2450	0.4870	0.000067778
2800	0.4950	0.000027778
3150	0.5090	0.000143333
3500	0.5150	0.000405556

RESULTS OF TEST
SIZE OF SAMPLES: N = 4000
USING THE UNIFORM DISTRIBUTION
WITH PARAMETERS A = 0.00D+00 B = 1.0
AVERAGE CPU TIME TAKEN OVER 10 RANDOM SAMPLES

VALUE OF M	AVERAGE CPU TIME	VARIANCE
400	0.6650	0.000294444
800	0.5780	0.000084444
1200	0.5620	0.000084444
1600	0.5530	0.000178889
2000	0.5600	0.000533333
2400	0.5690	0.000632222
2800	0.5640	0.000404444
3200	0.5550	0.000116667
3600	0.5680	0.000128889
4000	0.5690	0.000076667

Plot of average CPU time taken using uniform distribution over 10 random samples with $M=N$, $M=\frac{1}{2}N$ and $M=\frac{1}{10}N$ together with best 'least squares' straight lines through the origen

(Salford compiler)

Key:
- $\times$ $\alpha = \frac{1}{10}$
- $+$ $\alpha = \frac{1}{2}$
- $\circ$ $\alpha = 1$

x-axis: Sample size N (500, 1000, 1500, 2000, 2500, 3000, 3500, 4000)
y-axis: CPU time (.1 to .8)

Results from SORTTIME using the uniform distribution
with 10 random samples   (Salford compiler)
with fitted model $T(N, \alpha) = \left(\frac{\hat{A}}{\alpha} + \hat{B}\alpha + \hat{C}\right) N$

N = 2000

N = 1500

N = 1000

N = 500

average sorting →

α →

M = α N

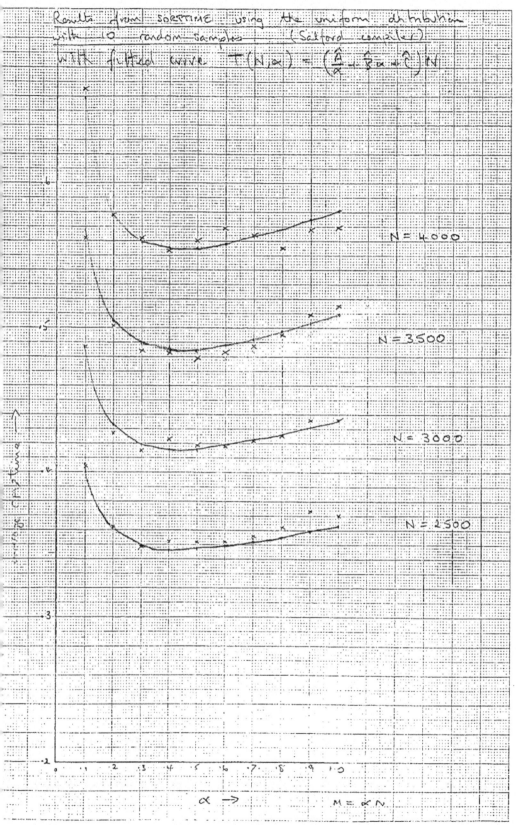

Results from SORTTIME using the uniform distribution
with 10 random samples. (Salford compiler)
with fitted curve $T(N, \alpha) = \left(\dfrac{\hat{A}}{\alpha} - \hat{B}\alpha + \hat{C} \right) N$

N = 4000

N = 3500

N = 3000

N = 2500

sorting time →

$\alpha \to$

$M = \alpha N$

Plot of variances from software using the uniform distribution with 10 random samples (Salford compiler)

α = 0.1
α = 0.5
α = 1.0

Variance s^2

3.3 Non-uniformly distributed samples.

We have run the program SORTTIME with three non-uniform distributions.
These are the standard normal, the exponential with mean one, and a U-shaped
distribution whose density function is given by $f(x) = \begin{cases} \frac{3}{2} x^2 & -1 \leqslant x \leqslant 1 \\ \\ 0 & \text{elsewhere} \end{cases}$

The results from these runs are plotted on pages **37**, **40** and **43**.

Examining the results using the standard normal distribution we see that as
with the uniform case we require a value of α between about 0.3 and 0.6 to
minimise the average CPU time elapsed; although for $N = 2000$ α should be
perhaps a little greater than 0.3 for a minimum. Again for $\alpha < 0.2$ the
elapsed time rises steeply. As we would expect the times themselves are a
little slower than for uniformly distributed samples. e.g. when $N = 1000$
the minimum is about 0.14 compared with 0.13 for the uniform case.

The results for the U-shaped distribution are similar to those for the
standard normal distribution with a value of α between 0.3 and 0.6 for a
minimum, and the times steeply rising for α less than 0.2.   The times themselves
are a little quicker than those for the standard normal distribution.

Using the exponential distribution with mean one a different pattern emerges.
Here we require α to be between 0.4 and 0.9 for a minimum, and the times
rise steeply for α less than 0.3.   However the minimum times themselves are
still only slightly higher than the uniform case. e.g. with $N = 1000$ the
minimum is about 0.15 compared with 0.13 for the uniform distribution.

In all cases increasing the value of α to 1.0 will only produce a small
increase in the average CPU time elapsed.

Obviously if we were to use a distribution whose density function had two
widely separated peaks the CPU time taken by the algorithm would increase
dramatically, and this method would be unsuitable. However we have seen that
provided α is large enough the method is robust and can withstand different
non-uniform distributions.

Results from SUBTIME using the standard normal distribution with 10 random samples   (Salford compiler)

N = 2000

N = 1500

N = 1500

N = 500

CPU Time (seconds)

0   .1   .2   .3   .4   .5   .6   .7   .8   .9   1.0

$\alpha \rightarrow$

$M = \alpha N$

-- 33 --

```
                    RESULTS OF TEST
SIZE OF SAMPLES: N =    500
USING THE   NORMAL      DISTRIBUTION
WITH PARAMETERS   A =   0.00D+00 B =    1.0
AVERAGE CPU TIME TAKEN OVER    10 RANDOM SAMPLES

VALUE OF M               AVERAGE CPU TIME

    50                      0.0900
   100                      0.0710
   150                      0.0660
   200                      0.0650
   250                      0.0660
   300                      0.0710
   350                      0.0630
   400                      0.0710
   450                      0.0690
   500                      0.0730


                    RESULTS OF TEST
SIZE OF SAMPLES: N =   1000
USING THE   NORMAL      DISTRIBUTION
WITH PARAMETERS   A =   0.00D+00 B =    1.0
AVERAGE CPU TIME TAKEN OVER    10 RANDOM SAMPLES

VALUE OF M               AVERAGE CPU TIME

   100                      0.1870
   200                      0.1430
   300                      0.1380
   400                      0.1400
   500                      0.1370
   600                      0.1410
   700                      0.1400
   800                      0.1440
   900                      0.1440
  1000                      0.1510
```

```
                    RESULTS OF TEST
SIZE OF SAMPLES: N =   1500
USING THE  NORMAL      DISTRIBUTION
WITH PARAMETERS  A =  0.00D+00 B =    1.0
AVERAGE CPU TIME TAKEN OVER    10 RANDOM SAMPLES

VALUE OF M              AVERAGE CPU TIME

    150                     0.2830
    300                     0.2310
    450                     0.2160
    600                     0.2150
    750                     0.2110
    900                     0.2100
   1050                     0.2140
   1200                     0.2180
   1350                     0.2210
   1500                     0.2270



                    RESULTS OF TEST
SIZE OF SAMPLES: N =   2000
USING THE  NORMAL      DISTRIBUTION
WITH PARAMETERS  A =  0.00D+00 B =    1.0
AVERAGE CPU TIME TAKEN OVER    10 RANDOM SAMPLES

VALUE OF M              AVERAGE CPU TIME

    200                     0.4260
    400                     0.3370
    600                     0.3180
    800                     0.3140
   1000                     0.3070
   1200                     0.3070
   1400                     0.3140
   1600                     0.3110
   1800                     0.3110
   2000                     0.3140
```

Results from sorttime using the exponential distribution
with mean 1 taken over ?0 random samples (sufficient...)

average CPU time

.4

.3

.2

.1

N = 2000

N = 1500

N = 1000

N = 500

0 ·1 ·2 ·3 ·4 ·5 ·6 ·7 ·8 ·9 1·0

α →

M ∝ N

— ka —

```
                   RESULTS OF TEST
SIZE OF SAMPLES: N =   500
USING THE  EXPONTL    DISTRIBUTION
WITH PARAMETERS  A =  0.00D+00 B =   1.0
AVERAGE CPU TIME TAKEN OVER    10 RANDOM SAMPLES

VALUE OF M            AVERAGE CPU TIME

    50                    0.1290
   100                    0.0930
   150                    0.0810
   200                    0.0780
   250                    0.0720
   300                    0.0760
   350                    0.0730
   400                    0.0750
   450                    0.0770
   500                    0.0750


                   RESULTS OF TEST
SIZE OF SAMPLES: N =  1000
USING THE  EXPONTL    DISTRIBUTION
WITH PARAMETERS  A =  0.00D+00 B =   1.0
AVERAGE CPU TIME TAKEN OVER    10 RANDOM SAMPLES

VALUE OF M            AVERAGE CPU TIME

   100                    0.2600
   200                    0.1850
   300                    0.1620
   400                    0.1510
   500                    0.1500
   600                    0.1510
   700                    0.1540
   800                    0.1590
   900                    0.1550
  1000                    0.1540
```

```
                    RESULTS OF TEST
SIZE OF SAMPLES: N =   1500
USING THE   EXPONTL     DISTRIBUTION
WITH PARAMETERS   A =   0.00D+00 B =    1.0
AVERAGE CPU TIME TAKEN OVER    10 RANDOM SAMPLES

VALUE OF M               AVERAGE CPU TIME

    150                      0.4040
    300                      0.3010
    450                      0.2670
    600                      0.2410
    750                      0.2360
    900                      0.2320
   1050                      0.2250
   1200                      0.2250
   1350                      0.2280
   1500                      0.2320


                    RESULTS OF TEST
SIZE OF SAMPLES: N =   2000
USING THE   EXPONTL     DISTRIBUTION
WITH PARAMETERS   A =   0.00D+00 B =    1.0
AVERAGE CPU TIME TAKEN OVER    10 RANDOM SAMPLES

VALUE OF M               AVERAGE CPU TIME

    200                      0.6990
    400                      0.4570
    600                      0.3940
    800                      0.3620
   1000                      0.3460
   1200                      0.3530
   1400                      0.3510
   1600                      0.3380
   1800                      0.3370
   2000                      0.3350
```

Results from SOFTTIME using a U-shaped distribution
with 10 random samples   (Salford compiler)

- average CPU time →
- x →
- M = α N
- N = 2000
- N = 1500
- N = 1000
- N = 500

- 412 -

```
                    RESULTS OF TEST
SIZE OF SAMPLES: N =    500
USING THE   U-SHAPE     DISTRIBUTION
WITH PARAMETERS  A =  0.00D+00 B =    1.0
AVERAGE CPU TIME TAKEN OVER    10 RANDOM SAMPLES

VALUE OF M               AVERAGE CPU TIME

     50                      0.0920
    100                      0.0730
    150                      0.0670
    200                      0.0650
    250                      0.0690
    300                      0.0650
    350                      0.0690
    400                      0.0690
    450                      0.0710
    500                      0.0720




                    RESULTS OF TEST
SIZE OF SAMPLES: N =   1000
USING THE   U-SHAPE     DISTRIBUTION
WITH PARAMETERS  A =  0.00D+00 B =    1.0
AVERAGE CPU TIME TAKEN OVER    10 RANDOM SAMPLES

VALUE OF M               AVERAGE CPU TIME

    100                      0.1820
    200                      0.1460
    300                      0.1370
    400                      0.1330
    500                      0.1320
    600                      0.1370
    700                      0.1370
    800                      0.1410
    900                      0.1420
   1000                      0.1440
```

OK, LF S5R6

RESULTS OF TEST

SIZE OF SAMPLES: N = 1500
USING THE U-SHAPE DISTRIBUTION
WITH PARAMETERS A = 0.00D+00 B = 1.0
AVERAGE CPU TIME TAKEN OVER 10 RANDOM SAMPLES

VALUE OF M	AVERAGE CPU TIME
150	0.2750
300	0.2270
450	0.2070
600	0.2080
750	0.2060
900	0.2050
1050	0.2090
1200	0.2110
1350	0.2110
1500	0.2130

RESULTS OF TEST

SIZE OF SAMPLES: N = 2000
USING THE U-SHAPE DISTRIBUTION
WITH PARAMETERS A = 0.00D+00 B = 1.0
AVERAGE CPU TIME TAKEN OVER 10 RANDOM SAMPLES

VALUE OF M	AVERAGE CPU TIME
200	0.4150
400	0.3310
600	0.3100
800	0.3000
1000	0.2980
1200	0.3000
1400	0.2940
1600	0.2990
1800	0.3030
2000	0.3050

3.4 Conclusion

We have seen in 2.4.2 that the address calculation sorting algorithm is most efficient when sorting a random sample from the uniform distribution. When this is the case the expected number of comparison statements executed will be $N(N-1)/4M + N$. Putting $M = \alpha N$ this becomes approximately $N/4\alpha + N$. So provided we take $M = \alpha N$ for some fixed α the algorithm is a method which is linear in N.

Comparing with an available number sorting routine in the NAG library, a routine which is a combination of Quicksort and Shellsort, we see that the address calculation algorithm is a fast sorting routine. That is when sorting 1000 uniformly distributed numbers the NAG routine takes about .22 seconds CPU time, whereas the address calculation routine takes about .13 seconds CPU time.

The major disadvantage of the address calculation routine is its use of additional storage space. It requires $N + 2M + 2$ extra integer elements and the sorted numbers must be placed in a new array of size N. This can be compared with the NAG sorting routine which requires 44 integer elements additional storage.

Finally in 3.3 we have seen that, provided (unusual) distributions for which the algorithm is specifically unsuited are avoided, the address calculation algorithm is robust to changes in the generating distribution.

The program SORTTIME.F77.

PROGRAM SORTA

```
C    A program to test the address calculation linked linear list insertionsort
C    algorithm.  Random samples of size  N  from a specified distribution are
C    generated and the average  CPU  time taken by the algorithm per sample is
C    calculated.  This is repeated for  NT  different values of  M,  the number
C    of intervals into which the range of the samples is divided; and for each
C    value of  M  the average  CPU  time is calculated from the results for  NRT
C    random samples.
        PARAMETER  ISIZE = 4000,  ILIST = 4001
        REAL  X(0:ISIZE),  W(0:ISIZE)
        INTEGER  P(0:ISIZE),  LP(ILIST),  L(ILIST)
        CHARACTER  NAME*10
        INTEGER*4  ITIME1,  ITIME2
        DOUBLE PRECISION  A, B, AVTM, A2TM, VARTM, TIME
C    The random sample is stored in  X(1:N).  The arrays  LP  and  P  contain
C    the initial pointers and pointers for the  M+1  lists.  The array  L  records
C    the number of elements in each list.
C    The size  N  of each sample, the number  NRT  of samples per trial, the
C    number  NT  of trials to be made, the distribution used  NAME, and its paramete
C    A  and  B,  are obtained from subroutine  DATA1.
        CALL  DATA1(N,NT,NRT,NAME,A,B)
C    Print page heading.
        WRITE(6,100) N,NAME,A,B,NRT
C    Begin testing.
        DO 90, IJ=1, NT
C    Calculate M.
        M = (N/NT)*IJ
```

```fortran
C     Begin a trial.
      A2TM = 0.0
      AVTM = 0.0
      DO 80, IK=1, NRT
C     Obtain a random sample from subroutine  GEN.
      CALL  GEN(N,IK,X,A,B)
C     Begin counting the  CPU  time.
      CALL  CTIM$A(ITIME1)
C     Find the range of  X.
      XMAX = X(1)
      XMIN = X(1)
      DO 10, I=2, N
      IF(X(I).GT.XMAX)THEN
      XMAX = X(I)
      ELSE
      IF(X(I).LT.XMIN) XMIN = X(I)
      END IF
   10 CONTINUE
      RNGE = XMAX - XMIN
C     Set  P,  LP;  and  L  to zero; set  X(0)  to infinity.
      DO 18, I=0, N
   18 P(I) = 0
      DO 19, I=1, M+1
      L(I) = 0
   19 LP(I) = 0
      X(0) = 1E38
C     Begin sorting  X.
      DO 50, J=1, N
```

```
C     Calculate the list  K  into which  X(J)  is to be fitted.
      K = ((X(J) - XMIN)/RNCE)*M
      K = K + 1
C     Calculate by accumulation the number of elements in each list.
      L(K) = L(K) + 1
C     Insert  X(J)  into list  K.
C     Is  X(J)  the smallest so far in list  K?
      IF(X(J).LE.X(LP(K))) THEN
      P(J) = LP(K)
      LP(K) = J
      ELSE
C     If  X(J)  is not the smallest element so far in list  K,  then compare  X(J
C     with each element of list  K  in turn.
      NEXT = LP(K)
   30 IF(X(J).LE.X(P(NEXT))) GOTO  31
      NEXT = P(NEXT)
      GOTO 30
   31 P(J) = P(NEXT)
      P(NEXT) = J
      END IF
   50 CONTINUE
C     Put the elements of  X  into  W  so that  W  is sorted.
      I = 1
      DO 60, K=1, M+1
      NEXT = LP(K)
      DO 60, J=1, L(K)
      W(I) = X(NEXT)
      NEXT = P(NEXT)
   60 I = I + 1
C     Put  X  equal to  W.
      DO 70, I=1, N
   70 X(I) = W(I)
```

```
C     Calculate the  CPU  time used.
      CALL CTIM$A(ITME2)
      TIME = ITIME2 - ITIME1
      AVTM = AVTM + TIME
      A2TM = A2TM + TIME*TIME
  80  CONTINUE
C     Write out the value of  M,  the average  CPU  time, and the variance.
      AVTM = AVTM/NRT
      AVTM = AVTM/100.00
      A2TM = A2TM/10000.00
      VARTM = (A2TM - NRT*AVTM*AVTM)/(NRT-1)
      WRITE(6,101) M, AVTM, VARTM
  90  CONTINUE
      STOP
 100  FORMAT(15X,'RESULTS OF TEST'/'SIZE OF SAMPLES: N =',I6/'USING THE'
     1     ,A12,' DISTRIBUTION'/'WITH PARAMETERS  A = ',G9.2,' B = ',
     2     G9.2/'AVERAGE CPU TIME TAKEN OVER',I6,' RANDOM SAMPLES'//
     3     'VALUE OF M',10X,'AVERAGE CPU TIME',10X,'VARIANCE'/)
 101  FORMAT(I8,20X,F7.4,10X,F14.9)
      END


      SUBROUTINE DATA1(M,NT,NRT,NAME,A,B)
      CHARACTER NAME*10
      DOUBLE PRECISION A,  B
      PRINT *, 'DATA PLEASE'
      PRINT *, 'SAMPLE SIZE  M = ?'
      READ *, N
      PRINT *, 'NUMBER OF SAMPLES PER TRIAL  NRT = ?'
      READ *, NRT
      PRINT *, 'NUMBER OF TRIALS  NT = ?'
      READ *, NT
```

```fortran
      PRINT *, 'DISTRIBUTION USED:  NAME = ?'
      READ *, NAME
      PRINT *, 'PARAMETERS:  A = ?  B = ?'
      READ *,  A, B
      RETURN
      END


      SUBROUTINE  GEN(N,IK,X,A,B)
      REAL  X(0:N)
      DOUBLE  PRECISION  A, B, G05DAF
      CALL  G05CBF(IK)
      DO 10, I=1, N
 10   X(I) = G05DAF(A,B)
      RETURN
      END
```

REFERENCES

(1) A.V. Aho, J.E. Hopcroft, and J.D. Ullman, "The Design and Analysis of
     Computer Algorithms", Addison-Wesley, 1974.

(2) E. Horowitz, and S. Sahni, "Fundamentals of Computer Algorithms",
     Pitman, 1979.

(3) E.J. Isaac, and R.C. Singleton, "Sorting by Address Calculation",
     Journal of Association for Computing Machinery Vol. 3, 1956, pp 169-174.

(4) D.E. Knuth, "The Art of Computer Programming Vol. 3, Sorting and Searching",
     Addison-Wesley, 1973.

(5) E.M. Reingold, J. Nievergelt, and N. Deo, "Combinatorial Algorithms
     Theory and Practice", Prentice-Hall, 1977.

(6) M.E. Tarter, and R.A. Kronmal, "Non-uniform key distribution and address
     calculation sorting", Proceedings Association for Computing Machinery
     National Conference Vol. 21, 1966, pp 331-337.

2. General Topology

Transfinite Extensions

Ordinal Invariants

TRANSFINITE EXTENSIONS

This paper follows the conventions of [**Thesis**], whereby all spaces are assumed to be completely regular Hausdorff, and for a space X, βX denotes the Stone-Čech compactification of X, and $X^* = \beta X \setminus X$ denotes the remainder of X in its Stone-Čech compactification.

Let $U = \{X_\alpha; f_{\alpha\beta}\}_{\alpha \in A}$ and $V = \{Y_\alpha; g_{\alpha\beta}\}_{\alpha \in A}$ be inverse limit systems indexed by the same directed set A. A map Φ from U to V is a collection $\{\varphi_\alpha: \alpha \in A\}$ of mappings φ_α indexed by the set A and such that, for each α, $\varphi_\alpha: X_\alpha \to Y_\alpha$, and for $\beta \le \alpha$ the diagram

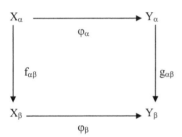

commutes.

Φ is said to be continuous, surjective, perfect, respectively if each of the mappings φ_α is continuous, surjective, perfect, respectively. (Note: A perfect mapping is defined to be continuous, closed, and having compact fibres (i.e. the inverse images of points are compact.)).

Let X_∞, Y_∞ be inverse limit spaces of U and V with canonical mappings f_α, g_α. The mapping $\varphi: X_\infty \to Y_\infty$ induced by Φ is defined by $\varphi((x_\alpha)_{\alpha \in A}) = (\varphi_\alpha(x_\alpha))_{\alpha \in A}$.

So for each α the diagram

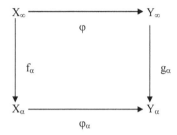

commutes.

Note also that φ is properly defined since if $\beta \leq \alpha$ then $f_{\alpha\beta}(x_\alpha) = x_\beta$ so $g_{\alpha\beta}(\varphi_\alpha(x_\alpha)) = \varphi_\beta(f_{\alpha\beta}(x_\alpha)) = \varphi_\beta(x_\beta)$.

Lemma 1.1

(i)   If Φ is continuous then φ is continuous.

(ii)   If Φ is perfect and surjective then φ is surjective

(iii)   If Φ is perfect and each of U and V are perfect systems the φ is perfect.

(iv)   If Φ is perfect and surjective and each of U and V are perfect systems then φ is a perfect surjection.

Proof

(i)   Since $g_\alpha \circ \varphi = f_\alpha \circ \varphi_\alpha$ for each α, if Φ is continuous then $g_\alpha \circ \varphi$ is continuous for each α and hence φ is continuous.

(ii)   Let $y = (y_\alpha)_{\alpha \in A}$ be an element of Y. Then for each α, $\varphi_\alpha^{-1}(y_\alpha)$ is compact since φ_α is perfect.

For each α put $Z_\alpha = \varphi_\alpha^{-1}(y_\alpha)$, so that Z_α is compact Hausdorff. Then if $\beta \leq \alpha$, $f_{\alpha\beta}(Z_\alpha) = f_{\alpha\beta}(\varphi_\alpha^{-1}(y_\alpha))$.

So $\varphi_\beta(f_{\alpha\beta}(Z_\alpha)) = \varphi_\beta \circ f_{\alpha\beta}(\varphi_\alpha^{-1}(y_\alpha)) = g_{\alpha\beta} \circ \varphi_\alpha(\varphi_\alpha^{-1}(y_\alpha)) = g_{\alpha\beta}(y_\alpha) = y_\beta$, and hence $f_{\alpha\beta}(Z_\alpha) \subseteq \varphi_\beta^{-1}(y_\beta) = Z_\beta$.

55

Then defining $h_{\alpha\beta}:Z_\alpha \to Z_\beta$ by $h_{\alpha\beta}(t) = f_{\alpha\beta}(t)$ for each t in Z_α we obtain an inverse limit system $W = \{Z_\alpha; h_{\alpha\beta}\}_{\alpha \in A}$ of non-empty compact Hausdorff spaces, since each φ_α is surjective.

Hence, by [**Thesis**] (7.1)(iii), the inverse limit space Z_∞ of W is non-empty. Let $z = (z_\alpha)_{\alpha \in A}$ be an element of Z_∞, then, clearly, it is also an element of X_∞, and, furthermore, by construction,

$g_\alpha \circ \varphi(z) = \varphi_\alpha \circ f_\alpha(z) = \varphi_\alpha(z_\alpha) = y_\alpha$ for each α.

Thus $\varphi(z) = (y_\alpha)_{\alpha \in A} = y$, and hence φ is surjective.

(iii)   Since U and V are perfect systems we know, by [**Thesis**] (7.2), that the canonical mappings f_α and g_α are also perfect.

Since Φ is continuous we know, by (i) that φ is continuous.

So if $y \in Y_\infty$ then $\varphi^{-1}(y)$ is a closed subspace of

$\varphi^{-1}(g_\alpha^{-1}(g_\alpha(y))) = (g_\alpha \circ \varphi)^{-1}(g_\alpha(y)) = (\varphi_\alpha \circ f_\alpha)^{-1}(g_\alpha(y))$, which is compact since $\varphi_\alpha \circ f_\alpha$ is a composition of perfect mappings and is therefore perfect.

Hence, $\varphi^{-1}(y)$ is compact, and so φ has compact fibres.

We now show that φ is a closed mapping.

Let F be a closed subspace of X_∞. Let $y \in Y \setminus \varphi(F)$.

Suppose, if possible, that $g_\alpha(y) \in \varphi_\alpha \circ f_\alpha(F)$ for each α.

Let $y_\alpha = g_\alpha(y)$ for each α.

The $\varphi_\alpha^{-1}(y_\alpha)$ is compact since φ_α is perfect; and, $f_\alpha(F)$ is closed since f_α is perfect.

Hence, $\varphi_\alpha^{-1}(y_\alpha) \cap f_\alpha(F)$ is compact and non-empty for each α.

Let $Z_\alpha = \varphi_\alpha^{-1}(y_\alpha) \cap f_\alpha(F)$.

Put $h_{\alpha\beta}$ equal to the restriction of $f_{\alpha\beta}$ to Z_α. Then as in the proof of (ii)

$W = \{Z_\alpha; h_{\alpha\beta}\}_{\alpha\in A}$ is an inverse limit system of non-empty compact Hausdorff spaces and so has a non-empty limit space Z_∞.

Let $z \in Z_\infty$ then, again as in (ii), z is an element of X_∞, and moreover,

$f_\alpha(z) \in f_\alpha(F)$ and $\varphi_\alpha \circ f_\alpha(z) = y_\alpha$ for each α.

Choose an element $x_{(\alpha)}$ if F for each α such that $f_\alpha(x_{(\alpha)}) = f_\alpha(z)$. Then for each δ, the net $(x_{(\alpha)})_\alpha$ is eventually inside the compact set

$(\varphi_\delta \circ f_\delta)^{-1}(y_\delta)$: for if $\alpha \geq \delta$ then

$(\varphi_\delta \circ f_\delta)(x_{(\alpha)}) = \varphi_\delta \circ (f_{\alpha\delta} \circ f_\alpha)(x_{(\alpha)}) = \varphi_\delta \circ f_{\alpha\delta}(f_\alpha(x_{(\alpha)})) = \varphi_\delta \circ f_{\alpha\delta}(f_\alpha(z)) =$

$\varphi_\delta \circ f_\delta(z) = y_\delta$.

Let x be a cluster point of the net $(x_{(\alpha)})_\alpha$.

Then $x \in (\varphi_\alpha \circ f_\alpha)^{-1}(y_\alpha)$ for each α; and further, since F is closed and $x_{(\alpha)} \in F$ for every α, $x \in F$.

Thus there is an x in F with $\varphi_\alpha(f_\alpha(x)) = y_\alpha$ for each α.

That is there is an x in F with $g_\alpha \circ \varphi(x) = y_\alpha$ for each α.

Then $\varphi(x) = (y_\alpha)_{\alpha\in A} = y$.

But $y \notin \varphi(F)$ so this is impossible.

So there must exist an α such that $g_\alpha(y) \notin \varphi_\alpha \circ f_\alpha(F)$.

For such an α choose a nbd U_α of $g_\alpha(y)$ in Y_α with U_α disjoint from the closed set $\varphi_\alpha \circ f_\alpha(F)$.

Then $g_\alpha^{-1}(U_\alpha)$ is an open nbd of y in Y_∞ and $g_\alpha^{-1}(U_\alpha)$ is disjoint form $\varphi(F)$: for if not there would be an x in F with $g_\alpha \circ \varphi(x) \in U_\alpha$, but

$g_\alpha \circ \varphi(x) = \varphi_\alpha \circ f_\alpha(x)$ and U_α is disjoint from $\varphi_\alpha \circ f_\alpha(F)$.

Hence φ is a closed mapping.

❩①❷▫This follows immediately from (ii) and (iii). ∎

Suppose now that for each space X there is a space FX and a mapping

$f_X:FX \to X$; and that for each space Y there is a space GY and a mapping

$g_Y:GY \to Y$.

Suppose also that for each space X and Y and for each continuous mapping

$h:X \to Y$ there is a uniquely determined continuous $h^*:FX \to GY$ which is such that the diagram

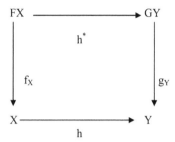

commutes.

For spaces X and Y let $\{X(\alpha)_F; f_{\alpha\beta}\}_\alpha$ and $\{Y(\alpha)_G; g_{\alpha\beta}\}_\alpha$ be the ordinal indexed classes defined as in [**Thesis**] (7.4).

Then we have

Lemma 1.2

Let the situation be as above. Then for each continuous mapping $h:X \rightarrow Y$ between spaces X and Y there is a unique ordinal indexed class $\{h(\alpha)\}_\alpha$ of continuous mappings $h(\alpha):X(\alpha)_F \rightarrow Y(\alpha)_G$ such that

(i) $h(0) = h$;

(ii)   for each ordinal α, $H(\leq \alpha) = \{h(\beta)\}_{\beta \leq \alpha}$ is a map between the inverse limit systems $U(\leq \alpha)_F = \{X(\beta)_F; f_{\beta\gamma}\}_{\beta \leq \alpha}$ and $V(\leq \alpha)_G = \{Y(\beta)_G; g_{\beta\gamma}\}_{\beta \leq \alpha}$;

(iii)   if $\alpha = \beta + 1$ then $h(\alpha) = h(\beta)^*$; and

(iv)   if α is a limit ordinal then $h(\alpha)$ is the mapping induced by the map $\{h(\beta)\}_{\beta < \alpha}$ between the inverse limit systems $U(< \alpha)_F = \{X(\beta)_F; f_{\beta\gamma}\}_{\beta < \alpha}$ and $V(< \alpha)_G = \{Y(\beta)_G; g_{\beta\gamma}\}_{\beta < \alpha}$.

Proof

To satisfy (i) we must take $h(0) = h$.

59

Let $\alpha > 0$ and suppose that for each $\beta < \alpha$ we have a continuous mapping

$h(\beta){:}X(\beta)_F \to Y(\beta)_G$ such that (i) to (iv) are satisfied for each $\beta < \alpha$.

Case 1: α a non-limit ordinal.

Let $\alpha = \beta + 1$. Then to satisfy (iii) we must take $h(\alpha) = h(\beta)^*$.

Then since $f_{\alpha\beta} = f_{F(\beta)X}$ and $g_{\alpha\beta} = g_{G(\beta)Y}$ and $X(\alpha)_F = FX(\beta)_F$ and $Y(\alpha)_G = GY(\beta)_G$ we see that the diagram

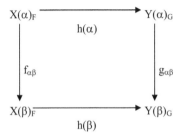

commutes; and since, by hypothesis, for each $\delta \leq \beta$ the diagram

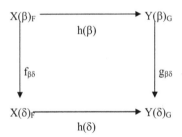

commutes, it is clear by taking composite mappings that (ii) is satisfied for α.

Case 2: α a limit ordinal.

If α is a limit ordinal then since (ii) holds for all $\beta < \alpha$, $H(< \alpha) = \{h(\beta)\}_{\beta < \alpha}$ is a map between the inverse limit systems $U(< \alpha)_F$ and $V(< \alpha)_G$.

Then (iv) we must take h(α) to be the mapping induced by the map H(< α).

Then certainly h(α):X(α)$_F$ → Y(α)$_G$.

Also for β < α, $f_{\alpha\beta}$ and $g_{\alpha\beta}$ are the canonical mappings and so the diagram

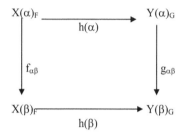

commutes.

So (ii) is satisfied for α.

This completes the induction. ∎

For spaces X and Y and h:X → Y a perfect mapping, as in [**Thesis**] p96, there is a uniquely determined perfect mapping h^1:X^* → Y^* which is surjective if h is surjective. So there is a uniquely determined h^2:X^{**} → Y^{**}, and h^2 is surjective if h is surjective.

Now, as in [**Thesis**] p108, let FX = X^{**} and $f_X = \bar{s}_{X,}$, GY = Y^{**} and $g_Y = \bar{s}_Y$, and denote the ordinal indexed classes determined by [**Thesis**] (7.4) be U(α) = {X(α); $s_{\alpha\beta}$}$_\alpha$ and V(α) = {Y(α); $t_{\alpha\beta}$}$_\alpha$.

Then for a perfect mapping h:X → Y we have a unique perfect mapping h^*:FX → GY where $h^* = h^2$.

Also the diagram

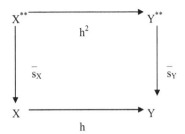

commutes, since both $\bar{s}_Y \circ h^2$ and $h \circ \bar{s}_X$ are, essentially, restrictions to X^{**} of the extension mapping $h^1:X^* \to \beta Y^*$ to βX^*.

So the diagram

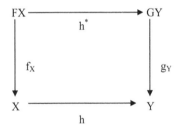

commutes.

Corollary 1.3

Let X and Y be spaces and let $U(\alpha) = \{X(\alpha); s_{\alpha\beta}\}_\alpha$ and $V(\alpha) = \{Y(\alpha); t_{\alpha\beta}\}_\alpha$ be as above. Let $h:X \to Y$ be a perfect mapping. Then there is a unique ordinal indexed class $\{h(\alpha)\}_\alpha$ of perfect mappings $h(\alpha):X(\alpha) \to Y(\alpha)$ such that conditions (i) to (iv) of lemma 1.2 are valid. Moreover, if h is surjective then $h(\alpha)$ is surjective for each α.

Proof

Since we know from the above that from perfect mappings k we obtain perfect mappings k^*, and , from [**Thesis**] (7.6), that each of $s_{\alpha\beta}$ and $t_{\alpha\beta}$ is perfect, and from

(1.1) (iii), that perfect maps between perfect limit systems induce perfect mappings, the uniqueness and existence of the class $\{h(\alpha)\}_\alpha$, although not a direct consequence of (1.2), is clear from the proof of (1.2).

Also, since perfect surjections k yield perfect surjections k^*, and, from (1.1) (iv) perfect surjective maps between perfect inverse limit systems induce perfect surjective mappings, it is again clear that each $h(\alpha)$ will be surjective if h is surjective.
■

For a space X let RX be the space R(X), the residue of X, (i.e. R(X) = {x ∈ X: x has no compact neighbourhood in X}), and let r_X:RX → X be the inclusion mapping. Then let $\{X(\alpha)_R; r_{\alpha\beta}\}_\alpha$ be the ordinal indexed class determined by R as in [**Thesis**] (7.4).

For a perfect mapping h:X → Y between spaces X and Y, let h^1:X^* → Y^* be the restriction of β(h), as above, and let h^*:X^{**} → Y be the restriction to X^{**} of the extension of h^1 to a mapping from βX^* into βY.

Then h^* is perfect, its range is contained in R(Y), and the diagram

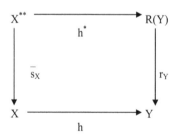

commutes, since both $r_Y \circ h^*$ and $h \circ \bar{s}_X$ are, essentially extensions of h^1.

Also, if h is surjective then as a mapping from X^{**} into $R(Y)$, h^* is also surjective.

Now since r_Y is perfect, it is clear, by the same reasoning as in [**Thesis**] (7.6), that each $r_{\alpha\beta}$ is a perfect mapping.

So just as we obtained (1.3) we also have

Corollary 1.4

Let X and Y be spaces. Let $U(\alpha) = \{X(\alpha); s_{\alpha\beta}\}_\alpha$ and $V(\alpha)_R = \{Y(\alpha)_R; r_{\alpha\beta}\}_\alpha$. Let $h:X \to Y$ be a perfect mapping. Then there is a unique ordinal indexed class $\{h(\alpha)\}_\alpha$ of perfect mappings $h(\alpha):X(\alpha) \to Y(\alpha)_R$ such that conditions (i) to (iv) of Lemma 1.2 are valid. Also, if h is surjective then so is $h(\alpha)$ for each α.

In particular if we take $Y = X$ and $h = id_X$ we obtain

Corollary 1.5

Let X be a space. Then there is a unique ordinal indexed class $\{id_X(\alpha)\}_\alpha$ of perfect surjections $id_X(\alpha):X(\alpha) \to X(\alpha)_R$ such that conditions (i) to (iv) of Lemma 1.2 are valid.

Let $\{Z_\alpha\}_{\alpha\in A}$ be a collection of spaces indexed by a directed set A and such that for each α in A, Z_α is a subspace of Z_β whenever $\alpha \geq \beta$.

For each α in A and for each $\beta \leq \alpha$, let $i_{\alpha\beta}:Z_\alpha \to Z_\beta$ be the inclusion mapping. Then, clearly, $W = \{Z_\alpha; i_{\alpha\beta}\}_{\alpha\in A}$ is an inverse limit system.

64

Let Z_∞ be the inverse limit space of W and $i_\alpha:Z_\infty \to Z_\alpha$ the canonical mappings for each α. Let $Z = \cap_{\alpha \in A} Z_\alpha$, and for each α in A let $k_\alpha:Z \to Z_\alpha$ be the inclusion mapping.

Lemma 1.6

There is a homeomorphism $h:Z_\infty \to Z$ such that for each α in A the diagram

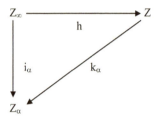

commutes.

Proof

Define $k:Z \to Z_\infty$ by $k(z) = (k_\alpha(z))_{\alpha \in A}$ for each z in Z. This is a valid definition since if $\alpha \in A$ and $\beta \leq \alpha$ then $k_\beta = i_{\alpha\beta} \circ k_\alpha$, so $i_{\alpha\beta}(k_\alpha(z)) = k_\beta(z)$.

Also, for each α, $k \circ i_\alpha = k_\alpha$ which is continuous, and so k is continuous.

Claim: If $z \in Z_\infty$ then $i_\alpha(z) = i_\beta(z)$ for each α and β in A.

To prove this, let α, $\beta \in A$ and choose a δ in a with with $\delta \geq \alpha$ and $\delta \geq \beta$. Then $i_\alpha(z) = i_{\delta\alpha} \circ i_\delta(z)$ which is the element $i_\delta(z)$; and similarly $i_\beta(z)$ is also the element $i_\delta(z)$.

Thus choose a fixed α_0 in A, and define $h:Z_\infty \to Z$ by $h(z) = i_{\alpha_0}(z)$. This definition is valid since our claim shows that $i_{\alpha_0}(z)$ must belong to Z_α for each α in A.

Also, clearly, the diagram

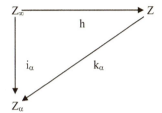

commutes for each α.

Then since $i_{\alpha 0}$ is continuous, from the definition, h must be continuous.

Finally, in view of our claim, it is clear that $k \circ h = id_{Z\infty}$ and $h \circ k = id_Z$ and so h is a homeomorphism with the required properties. ■

Lemma 1.7

Let $\Phi = \{\varphi_\alpha\}_{\alpha \in \Lambda}$ be a map between inverse limit systems $U = \{X_\alpha; f_{\alpha\beta}\}_{\alpha \in \Lambda}$ and $V = \{Y_\alpha; g_{\alpha\beta}\}_{\alpha \in \Lambda}$. Then if each φ_α is a homeomorphism, the induced mapping $\varphi:X_\infty \to Y_\infty$ is also a homeomorphism.

Proof

Let $\Psi = \{\varphi_\alpha^{-1}\}_{\alpha \in \Lambda}$. Then for each α and $\beta \leq \alpha$, $g_{\alpha\beta} \circ \varphi_\alpha = \varphi_\beta \circ f_{\alpha\beta}$, and so $\varphi_\beta^{-1} \circ g_{\alpha\beta} \circ \varphi_\alpha \circ \varphi_\alpha^{-1} = \varphi_\beta^{-1} \circ \varphi_\beta \circ f_{\alpha\beta} \circ \varphi_\alpha^{-1}$, that is $f_{\alpha\beta} \circ \varphi_\alpha^{-1} = \varphi_\beta^{-1} \circ g_{\alpha\beta}$; hence Ψ is a map from U to V.

Let $\psi:Y_\infty \to X_\infty$ be the mapping induced by Ψ.

Let $y = (y_\alpha)_{\alpha \in \Lambda}$ be an element of Y_∞.

Then $\varphi \circ \psi(y) = \varphi(\psi(y)) = \varphi((\varphi_\alpha^{-1}(y_\alpha))_{\alpha \in \Lambda}) = (\varphi_\alpha(\varphi_\alpha^{-1}(y_\alpha)))_{\alpha \in \Lambda} = (y_\alpha)_{\alpha \in \Lambda} = y$.

So $\varphi \circ \psi = id_{Y\infty}$, and similarly $\psi \circ \varphi = id_{X\infty}$.

Hence φ is a homeomorphism. ■

For a space X and ordinals α, β with $\alpha \geq \beta$ let $i_{\alpha\beta}:R^{\alpha}(X) \to R^{\beta}(X)$ be the inclusion mapping. Then, clearly, the class $W = \{R^{\alpha}(X); i_{\alpha\beta}\}_{\alpha}$ is such that for each ordinal α, $W(\leq \alpha) = \{R^{\beta}(X); i_{\beta\gamma}\}_{\beta\leq\alpha}$ is an inverse limit system.

Lemma 1.8

Let X be a space and α a limit ordinal. Suppose that we have an inverse limit system $V(<\alpha) = \{J_{\beta}; j_{\beta\gamma}\}_{\beta<\alpha}$ indexed by the set of ordinals less than α; and a map $\Phi = \{\varphi_{\beta}\}_{\beta<\alpha}$ from $V(< \alpha) = \{J_{\beta}; j_{\beta\gamma}\}_{\beta<\alpha}$ to $W(< \alpha) = \{R^{\beta}(X); i_{\beta\gamma}\}_{\beta<\alpha}$ such that each φ_{β} is a homeomorphism. Then there is a homeomorphism $\psi:J_{\infty} \to R^{\alpha}(X)$ such that for each β the diagram

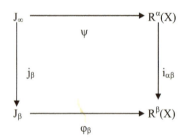

commutes.

Proof

Let R_{∞} be the inverse limit space of $W(< \alpha)$, and for each β, $i_{\beta}:R_{\infty} \to R^{\beta}(X)$ be the canonical mapping.

Let $\varphi:J_{\infty} \to R_{\infty}$ be the mapping induced by Φ.

Then the diagram

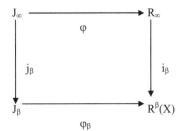

commutes for each β, and, by (1.7), φ is a homeomorphism.

Let $h : R_\infty \to R^\alpha(X)$ be the homeomorphism obtained as in Lemma 1.6.

Then, by (1.6), the diagram

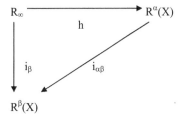

commutes for each β.

Hence $\psi = h \circ \varphi$ is a homeomorphism with the required properties. ∎

Lemma 1.9

Let X be a space. Let $U(\alpha)_R = \{X(\alpha)_R; r_{\alpha\beta}\}_\alpha$ and $W(\alpha) = \{R^\alpha(X); i_{\alpha\beta}\}_\alpha$ be the ordinal indexed classes defined above. Then there is a class $\Phi(\alpha) = \{h_\alpha\}_\alpha$ of homeomorphisms $h_\alpha : X(\alpha)_R \to R^\alpha(X)$ such that, for each α, $\Phi(\leq \alpha) = \{h_\beta\}_{\beta \leq \alpha}$ is a map from $U(\leq \alpha)_R$ to $W(\leq \alpha)$.

Proof

We define the class $\{h_\alpha\}_\alpha$ by induction.

Let $h_0 = id_X$.

Suppose that $\alpha > 0$ and we have homeomorphisms $\{h_\beta\}_{\beta<\alpha}$ with the required properties.

Case 1: α a non-limit ordinal.

Let $\alpha = \beta + 1$ then since $h_\beta : X(\beta)_R \to R^\beta(X)$ is a homeomorphism, in view of [**Thesis**] (5.3), the restriction of h_β to $R(X(\beta)_R)$, which equals $X(\alpha)_R$, determines a homeomorphism h_α from $X(\alpha)_R$ onto $R(R_\beta(X))$, which equals $R_\alpha(X)$.

Then it is clear that the collection $\{h_\beta\}_{\beta\leq\alpha}$ has the required properties.

Case 2: α a limit ordinal.

Let α be a limit ordinal.

Then using Lemma 1.8 there is a homeomorphism $\psi : X(\alpha)_R \to R^\alpha(X)$ such that the diagram

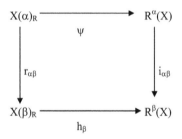

commutes for each $\beta \leq \alpha$.

Put $h_\alpha = \psi$. Then, clearly, the collection $\{h_\beta\}_{\beta\leq\alpha}$ has the required properties.

This completes the induction. ∎

We now combine (1.9) and (1.5) to obtain

<u>Theorem 1.10</u>

Let X be a space. Let $\{X(\alpha); s_{\alpha\beta}\}_\alpha$ and $\{R^\alpha(X); i_{\alpha\beta}\}_\alpha$ be as defined above. Then there is a class $\Phi(\alpha) = \{\varphi_\alpha\}_\alpha$ of perfect surjections $\varphi_\alpha:X(\alpha) \to R^\alpha(X)$ such that whenever $\beta \leq \alpha$ the diagram

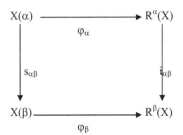

commutes; also $\varphi_0 = id_X$.

<u>Proof</u>

Let $\{id_X(\alpha)\}_\alpha$ be the class of perfect surjections obtained in (1.5), and $\{h_\alpha\}_\alpha$ be the class of homeomorphisms obtained in (1.9).

Then $id_X(0) = id_X$, $h_0 = id_X$, and for $\beta \leq \alpha$ we have the commutative diagram

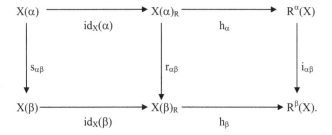

Hence taking φ_α to be $h_\alpha \circ id_X(\alpha)$ for each α we obtain a class $\Phi(\alpha) = \{\varphi_\alpha\}_\alpha$ of perfect surjections with the required properties. ∎

Corollary 1.11

Let X be a space. Then for each ordinal α, $s_{\alpha 0}:X(\alpha) \to X$ is a perfect mapping with range $R^\alpha(X)$.

Proof

Let α be an ordinal number. We know that $s_{\alpha 0}$ is perfect.

Let $\varphi_\alpha:X(\alpha) \to R^\alpha(X)$ be the perfect surjection provided by (1.10).

Then, by (1.10), the diagram

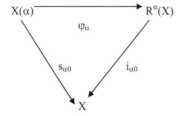

commutes.

Hence $s_{\alpha 0}(X(\alpha)) = i_{\alpha 0} \circ \varphi_\alpha(X(\alpha)) = i_{\alpha 0}(R^\alpha(X)) = R^\alpha(X)$. ∎

Let $U = \{X_\alpha; f_{\alpha\beta}\}_{\alpha \in A}$ be an inverse limit system. Let E be a cofinal subset of A. Then necessarily E is a directed set with the partial order inherited from A. Hence $U_E = \{X_\alpha; f_{\alpha\beta}\}_{\alpha \in E}$ is also an inverse limit system.

Let X_∞ be the inverse limit space of U and, for $\alpha \in A$, $f_\alpha:X_\infty \to X_\alpha$ the canonical mappings; and let Z_∞ be the inverse limit space of U_E and, for $\alpha \in E$, $g_\alpha:Z_\infty \to X_\alpha$ the canonical mappings.

Lemma 1.12

There is a homeomorphism $h:Z_\infty \to X_\infty$ such that for each α in E the diagram

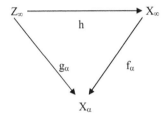

commutes.

Proof

Define $k:X_\infty \to Z_\infty$ by $k(x) = k((x_\alpha)_{\alpha \in A}) = (x_\alpha)_{\alpha \in E}$, where $x=(x_\alpha)_{\alpha \in A}$ is an element of X_∞.

Then, for each $\alpha \in E$, $g_\alpha \circ k = f_\alpha$, and so k must be continuous.

Now let $z = (z_\alpha)_{\alpha \in E}$ be an element of Z_∞. Define h(z) as follows: if $\alpha \in A$ choose a β in E with $\beta \geq \alpha$ and then take x_α in X_α to be equal to $f_{\beta\alpha}(z_\beta)$; and put $h(z) = (x_\alpha)_{\alpha \in A}$.

We show that this is a valid definition.

Firstly, for α in A suppose that both β and γ belong to E and $\beta \geq \alpha$ and $\gamma \geq \alpha$. Then there is a δ in E with $\delta \geq \beta$ and $\delta \geq \gamma$. Then $f_{\gamma\alpha} \circ g_\gamma(z) = f_{\gamma\alpha} \circ (f_{\delta\gamma} \circ g_\delta)(z) = f_{\delta\alpha} \circ g_\delta(z)$; and similarly, $f_{\beta\alpha} \circ g_\beta(z) = f_{\delta\alpha} \circ g_\delta(z)$.

So each x_α is well defined.

Secondly, if α and β belong to A with $\alpha \geq \beta$, then choose a γ in E with $\gamma \geq \alpha$, and $\gamma \geq \beta$. Then $f_{\alpha\beta}(x_\alpha) = f_{\alpha\beta}(f_{\gamma\alpha}(z_\gamma)) = f_{\gamma\beta}(z_\gamma) = x_\beta$; and so h(z) is an element of X_∞.

Also, if $\alpha \in E$ then $x_\alpha = z_\alpha$ and so the diagram

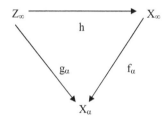

commutes for each α in E.

Let $\beta \in A$. Then there is an α in E with $\alpha \geq \beta$. Then

$f_\beta \circ h = (f_{\alpha\beta} \circ f_\alpha) \circ h = f_{\alpha\beta} \circ (f_\alpha \circ h) = f_{\alpha\beta} \circ g_\alpha$; so $f_\beta \circ h$ is continuous for each β in A, and hence h is continuous.

Finally, if $z = (z_\alpha)_{\alpha \in E}$ is an element of Z_∞, then for each α in E,

$g_\alpha \circ (k \circ h)(z) = (g_\alpha \circ k) \circ h(z) = f_\alpha \circ h(z) = g_\alpha(z)$; hence $k \circ h = id_{Z_\infty}$.

Also, if $x = (x_\alpha)_{\alpha \in A}$ is an element of X_∞, then, for each α in A, choose a β in E with $\beta \geq \alpha$, then

$f_\alpha \circ (h \circ k)(x) = (f_\alpha \circ h) \circ k((x_\alpha)_{\alpha \in A}) = f_\alpha \circ h((x_\alpha)_{\alpha \in E}) = f_{\beta\alpha}(x_\beta)$ (by definition of h)

$= x_\alpha = f_\alpha(x)$; and hence $h \circ k = id_{X_\infty}$.

It follows that h is a homeomorphism. ∎

<u>Remark 1.13</u>

Let h:X → Y be a homeomorphism between spaces X and Y. Then the extension mapping $\beta(h):\beta X \to \beta Y$ must also be a homeomorphism, and so $h^1:X^* \to Y^*$ is a homeomorphism. It follows that the mapping $h^2:X^{**} \to Y^{**}$ is again a homeomorphism. Also as we observed prior to (1.3) the diagram

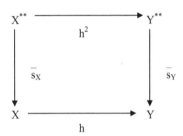

commutes.

Let now X be a space. Then we may form the class $\{X(\alpha); s_{\alpha\beta}\}_\alpha$. Let δ be an ordinal number, then from the space $X(\delta)$ we may likewise form the class $\{X(\delta)(\alpha); t_{\alpha\beta}\}_\alpha$.

<u>Lemma 1.14</u>

Let X be a space and δ an ordinal number. Then there is a class $\{h_\alpha\}_\alpha$ of homeomorphisms $h_\alpha:X(\delta)(\alpha) \to X(\delta + \alpha)$ such that whenever $\beta \le \alpha$ the diagram

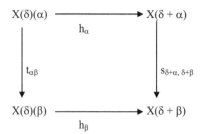

commutes; and $h_0 = id_{X(\delta)}$.

Proof

Let $h_0 = id_{X(\delta)}$.

assume that $\alpha > 0$ and that for each $\beta < \alpha$ we have defined a homeomorphism

$h_\beta : X(\delta)(\beta) \to X(\delta + \beta)$ such that whenever $\gamma \le \beta$ the diagram

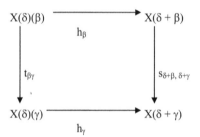

commutes.

Then we define h_α as follows:

If $\alpha = \beta + 1$, then from the homeomorphism $h_\beta : X(\delta)(\beta) \to X(\delta + \beta)$ we obtain,

by (1.13), a homeomorphism $h^2_\beta : X(\delta)(\beta)^{**} \to X(\delta + \beta)^{**}$.

But $X(\delta)(\beta)^{**} = X(\delta)(\alpha)$ and $X(\delta + \beta)^{**} = X(\delta + \alpha)$, so we can define a

homeomorphism h_α by putting $h_\alpha = h^2_\beta$, and then, by (1.13), the diagram

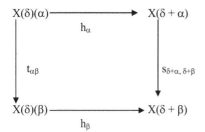

commutes.

Then, clearly, by composing mappings the diagram

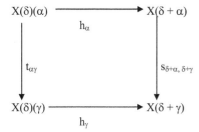

commutes for each $\gamma \leq \alpha$.

Now suppose that α is a limit ordinal.

We note from [**Thesis**] (1.12) that $\delta + \alpha$ is also a limit ordinal, and the set $\{\delta + \tau : \tau < \alpha\}$ is equal to the set $\{\kappa : \delta \leq \kappa < \delta + \alpha\}$.

Hence using the collection $\{h_\beta\}_{\beta < \alpha}$ of homeomorphisms we can identify the inverse limit system $\{X(\delta)(\beta); t_{\beta\gamma}\}_{\beta < \alpha}$ with the cofinal subsystem $\{X(\delta + \beta); s_{\delta+\beta, \delta+\gamma}\}_{\beta < \alpha}$ of the inverse limit system $\{X(\beta); s_{\beta\gamma}\}_{\beta < \delta + \alpha}$.

Hence we may apply Lemma 1.12 to obtain a homeomorphism $h_\alpha : X(\delta)(\alpha) \rightarrow X(\delta + \alpha)$ such that for each $\gamma \leq \alpha$ the diagram

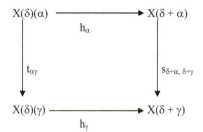

commutes.

This completes the inductive definition. ∎

Theorem 1.15

Let X be a space and α an ordinal number. Then for each $\delta \leq \alpha$,

$s_{\alpha\delta}: X(\alpha) \to X(\delta)$ is a perfect mapping with range $R^{\alpha-\delta}(X(\delta))$.

Proof

Let α be an ordinal and $\delta \leq \alpha$.

We already know that $s_{\alpha\delta}$ is a perfect mapping.

Now, by (1.14) there is a homeomorphism $h: X(\delta)(\alpha - \delta) \to X(\alpha)$ such that the

diagram

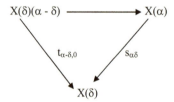

commutes.

Also, by (1.11), $t_{\alpha-\delta,0}(X(\delta)(\alpha - \delta)) = R^{\alpha-\delta}(X(\delta))$.

Hence $s_{\alpha\delta}(X(\alpha)) = R^{\alpha-\delta}(X(\delta))$. ∎

References

[**Thesis**]   "Iterated Remainders in Compactifications" by P. P. Jackson.

For further references see [**Thesis**].

ORDINAL INVARIANTS

For a, (completely regular Hausdorff), space X we know, by considering cardinals, that the ordinal-residues, $R^{\alpha}(X)$, must eventually be nowhere locally compact; in fact, they are all eventually equal to the largest closed nowhere locally compact subspace of X, which we have labelled $R^{\infty}(X)$.

Now also, for a space X, we know, from [**Thesis**] (7.22), that the ordinal indexed class $\{X(\alpha)\}_{\alpha}$ of "transfinite remainders" eventually consists of nowhere locally compact spaces.

Thus we are able to define the following ordinal invariants:

Definition 2.1

Let X be a space. Then

(i) $\sigma(X)$ shall denote the least ordinal α such that $R^{\alpha}(X)$ is nowhere locally compact;

(ii) $\tau(X)$ shall denote the least ordinal α such that $X(\alpha)$ is nowhere locally compact.

Example 2.2

We show that these invariants we have defined are not necessarily equal to one another.

Let X be the space of [**Thesis**] (6.10(ii)). Then as shown there, R(X) is nowhere locally compact, whilst X^{**}, (where X^* denotes the remainder of X in its Stone-Čech compactification βX, i.e. βX \ X), fails to be nowhere locally compact. Thus $\sigma(X) = 1$ whilst $\tau(X) > 1$.

They may however be equal since, for example, if Y is any nowhere locally compact space then both $\sigma(X)$ and $\tau(X)$ are zero.

Lemma 2.3

Let X be a space. Then, $\sigma(X) \leq \tau(X)$.

Proof

By [**Trans. Ext.**] (1.11) there is a perfect mapping from $X(\alpha)$ onto $R^\alpha(X)$ for each ordinal α. Hence, by [**Thesis**] (5.4)(ii), if $X(\alpha)$ is nowhere locally compact then $R^\alpha(X)$ must also be nowhere locally compact.

Thus the result is now clear from the definition of $\sigma(X)$ and $\tau(X)$. ∎

Lemma 2.4

Let f:X → Y be a perfect surjection. Then

(i) $\sigma(Y) \leq \sigma(X)$; and

(ii) $\tau(Y) \leq \tau(X)$.

81

Proof

(i)   For each α the restriction of f to $R^{\alpha}(X)$ defines, by [**Thesis**] (5.3)(ii), a perfect mapping from $R^{\alpha}(X)$ onto $R^{\alpha}(Y)$. Hence, if $R^{\alpha}(X)$ is nowhere locally compact then so is $R^{\alpha}(Y)$. Thus (i) follows from the definition of σ.

)◎◎◻For each α there is, as in (1.3), a perfect mapping, f(α), from X(α) onto Y(α). Hence, if X(α) is nowhere locally compact then so is Y(α). Thus (ii) follows from the definition of τ. ∎

Examples 2.5

We show that the existence of a perfect mapping from a space X onto a space Y neither implies $\sigma(X) = \sigma(Y)$, nor $\tau(X) = \tau(Y)$.

(i)   Let, as in [**Thesis**] (3.17)(ii), the space X be $D \cup E_Q$, the open unit disc with 'rational circumference'. Then there is, of course, a perfect mapping from X^*, the remainder of X in its Stone-Čech Compactification βX, onto $\overline{D} \setminus X$, the remainder of X in the closed unit disc $\overline{D}$.

Now, $\overline{D} \setminus X$ is, of course, nowhere locally compact and so $\sigma(\overline{D} \setminus X) = 0$. However, as shown in [**Thesis**] (3.17)(ii), X^* fails to be nowhere locally compact and so $\sigma(X^*) \neq 0$.

Then also, $\tau(\overline{D} \setminus X) = 0$ and $\tau(X^*) \neq 0$.

(ii)   Alternatively, let X be as in (2.2). Then R(X) is nowhere locally compact whilst X^{**} fails to be nowhere locally compact. But there is, of course, a perfect mapping, $\overline{s}_X$, from X^{**} onto R(X).

Thus $\sigma(X^{**}) \neq 0$, $\sigma(R(X)) = 0$, and $\tau(X^{**}) \neq 0$, $\tau(R(X)) = 0$.

82

We see, from (2.5) (i) above, that if αX and γX are compactifications of a space X, then for the remainders $\alpha X \setminus X$ and $\gamma X \setminus X$ it need not be true that the ordinal invariants $\sigma(\alpha X \setminus X)$ and $\sigma(\gamma X \setminus X)$ are equal. We can show, however, that they must be near each other.

Lemma 2.6

Let X be a space and αX and γX be compactifications of X. Then

(i) $\sigma(\gamma X \setminus X) \leq \sigma(\alpha X \setminus X) + 1$; and

(ii) $\sigma(\alpha X \setminus X) \leq \sigma(\gamma X \setminus X) + 1$.

Proof

We prove (i), and then (ii) must clearly follow from the symmetry of the situation.

Let $\sigma = \sigma(\gamma X \setminus X)$.

Then $R^{\sigma}(\gamma X \setminus X)$ is a nowhere locally compact space.

Hence, as defined in [**Thesis**] (7.12), the space

$S^{\sigma}(X) = cl_{\gamma X}(R^{\sigma}(\gamma X \setminus X)) \setminus R^{\sigma}(\gamma X \setminus X)$ is a remainder of a nowhere locally compact space and is therefore nowhere locally compact.

Then, again, $cl_{\alpha X}(S^{\sigma}(X)) \setminus S^{\sigma}(X)$ is a remainder of a nowhere locally compact space and is therefore nowhere locally compact.

But, by [**Thesis**] (7.11), $S^{\sigma}(X) \cup R^{\sigma}(\alpha X \setminus X)$ is a compactification of $R^{\sigma}(\alpha X \setminus X)$, and so, $cl_{\alpha X}(S^{\sigma}(X)) \setminus S^{\sigma}(X)$ is equal to $R^{\sigma+1}(\alpha X \setminus X)$.

Thus $R^{\sigma+1}(\alpha X \setminus X)$ is nowhere locally compact.

Hence $\sigma(\alpha X \setminus X) \leq \sigma + 1 = \sigma(\gamma X \setminus X) + 1$. ∎

<u>Example 2.7</u>

Let X be any non-compact space. Let P be the product of an infinite number of copies of X. Let diag(P) be the diagonal of P. Then diag(P) is a closed subspace of P and diag(P) is homeomorphic to X.

Now, by [**Thesis**] (4.24), P is a nowhere locally compact space, and so $\sigma(P) = 0$.

Also, for each α, $R^{\alpha}(P \oplus diag(P)) = R^{\alpha}(P) \oplus R^{\alpha}(diag(P)) = P \oplus R^{\alpha}(diag(P)) = P \oplus R^{\alpha}(X)$.

Hence $\sigma(P \oplus diag(P)) = \sigma(X)$.

However, we see that P is a perfect image of $P \oplus diag(P)$ since the mapping id_P and the inclusion diag(P) $\to$ P combine to form a perfect mapping, identity $\oplus$ inclusion, from $P \oplus diag(P)$ onto P.

We have seen in (2.4) that if Z is a perfect image of a space Y then $\sigma(Z) \leq \sigma(Y)$.

We have seen in (2.6) that if Z and Y are both remainders of the same space X, then, $\sigma(Y) \leq \sigma(Z) + 1$.

We have also seen, in (2.7), that the existence of a perfect surjection from Y onto Z does not in itself imply that $\sigma(Y) \leq \sigma(Z) + 1$; in fact, we may choose our space so that $\sigma(Y)$ can be as far from $\sigma(Z)$ as we please.

Now, for a locally compact space X, Magill's Theorem (1966) characterises the remainders of X as precisely the perfect images of X^{*}. Thus we have indirectly shown that the hypothesis of local compactness is necessary for the conclusions of Magill's Theorem to hold: For, from (2.7), we can obtain spaces Y and Z with $\sigma(Y)$

= 2 and $\sigma(Z) = 0$, and Z a perfect image of Y. Then, by (2.6), Y and Z cannot both be remainders of the same space; and, of course, there is, as in [**Thesis**] (2.7), some space X for which X^* is homeomorphic to Y.

We can say also that if Y is a remainder of a space X then necessarily Y is a perfect image of X^* and so, by (2.4), $\sigma(Y) \leq \sigma(X^*)$; and as both X^* and Y are remainders of X, by (2.6), $\sigma(X^*) \leq \sigma(Y) + 1$. So if Y is a remainder of a space X the $\sigma(Y)$ is either equal to or is the immediate predecessor of $\sigma(X^*)$. Further the condition $\sigma(X^*) \leq \sigma(Y) + 1$ is <u>not</u> implied by the existence of a perfect mapping from X^* onto Y.

Thus any attempt to extend Magill's Theorem to include spaces which are not necessarily locally compact, by using hypotheses additional to the requirement 'Y is a perfect image of X^*' must employ conditions which are sufficient to ensure that $\sigma(X^*) \leq \sigma(Y) + 1$.

Example 2.8

We show that a space Y may be a perfect image of X^* and satisfy $\sigma(X^*) \leq \sigma(Y) + 1$, and yet fail to be a remainder of X in any compactification.

Let $W = W(\omega_1)$ be the space of ordinals less than the first uncountable ordinal ω_1, and let $\overline{W} = W(\omega_1 + 1)$ be the space of ordinals less than or equal to the first uncountable ordinal. Then $\overline{W}$ is the one-point, Stone-Čech, (and only), compactification of W.

Let [0, 1] be the closed unit interval in the real line.

Then the projection mapping $p:W \times [0, 1] \to W$, is a projection parallel to a compact factor and is therefore perfect.

Now let $\aleph_\alpha$ be a cardinal greater than the cardinal of $\overline{W} \times [0, 1]$ and such that α is a non-zero, non-limit ordinal. Then, from [**Thesis**] (2.6),

$W(\omega_\alpha + 1) \times \overline{W} \times [0, 1]$ is the Stone-Čech compactification of $W(\omega_\alpha) \times \overline{W} \times [0, 1]$.

Let $X = W(\omega_\alpha + 1) \times \overline{W} \times [0, 1] \setminus \{\omega_\alpha\} \times W \times [0, 1]$.

Then $\beta X = W(\omega_\alpha + 1) \times \overline{W} \times [0, 1]$ and so,

$X^* = \beta X \setminus X = \{\omega_\alpha\} \times W \times [0, 1]$, which is homeomorphic to $W \times [0, 1]$.

Thus W is a perfect image of X^*.

Also, since both X^* and W are locally compact, $\sigma(X^*) = 1 = \sigma(W)$.

We now show that W cannot be a remainder of X in any compactification.

We have, $R(X) = cl_{\beta X}(X^*) \setminus X^* = \{\omega_\alpha\} \times \{\omega_1\} \times [0, 1]$.

But suppose, if possible, that W is a remainder of X. Then, necessarily, R(X) would be a remainder of W; and we know that the only remainder of W is the singleton space. Hence W is not a remainder of X in any compactification.

Lemma 2.9

Let X and Y be spaces. Suppose that R(X) is a remainder of Y. Then $\sigma(X^*) \le \sigma(Y) + 1$.

Proof

If R(X) is a remainder of Y then, from [**Thesis**] (3.6), both R(Y) and $R(X^*)$ are remainders of R(X).

Hence, by (2.6), $\sigma(R(X^*)) \le \sigma(R(Y)) + 1$.   (*)

Case 1: $\sigma(Y) = 0$.

If Y is nowhere locally compact then so is its remainder, R(X), and hence also the remainder $R(X^*)$ of R(X). Thus $\sigma(Y) = 0$ implies that $\sigma(X^*) \le 1$.

Case 2: $\sigma(Y)$ finite and non-zero.

Let $\sigma(Y) = n$ where n is finite and non-zero.

Then $R^n(Y)$ is nowhere locally compact.

Hence $R^{n+1}(X) = R^n(R(X))$, being a remainder of $R^n(Y)$, is also nowhere locally compact.

The $R^{n+1}(X^*)$, being a remainder of $R^{n+1}(X)$, is nowhere locally compact.

Thus $\sigma(X^*) \leq n + 1 = \sigma(Y) + 1$.

Case 3: $\sigma(Y)$ infinite.

Let $\sigma(Y)$ be infinite.

If $\sigma(X^*)$ is finite then the inequality $\sigma(X^*) \leq \sigma(Y) + 1$ obviously holds.

Suppose then that both $\sigma(Y)$ and $\sigma(X^*)$ are infinite.

Then, clearly, $\sigma(Y) = \sigma(R(Y))$ and $\sigma(X^*) = \sigma(R(X^*))$.

Hence, by (*), $\sigma(X^*) \leq \sigma(Y) + 1$. ■

Remark 2.10

If Y is a remainder of X then, necessarily, $R(X)$ is a remainder of Y. So the condition '$R(X)$ is a remainder of Y' is necessary for Y to be a remainder of X; and it implies our necessary condition of $\sigma(X^*) \leq \sigma(Y) + 1$.

Proposition 2.11

If $R^\sigma(X)$ is non-empty and nowhere locally compact then $X(\sigma)$ must fail to be locally compact.

We have a perfect surjection $X(\sigma) \to R^{\sigma}(X)$. Since local compactness is a fitting property, if $X(\sigma)$ were to be locally compact, then $R^{\sigma}(X)$ would also be locally compact, which is not the case.

Alternatively, if $X(\sigma)$ is locally compact, the $X(\sigma+1)$ is empty. But since $X(\sigma+1) \to R^{\sigma+1}(X)$, $[= R(^{\sigma}(X))]$, is surjective this is not possible. ∎

Definition 2.12

For a space X, the subspace $R^{\infty}(X)$, which we have alternatively characterised as the largest closed nowhere locally compact subspace of X, will be called the core of X.

Proposition 2.13

Let X be a space. If the core of X is empty the $\sigma(X) = \tau(X)$.

Proof

Let the core of X be empty. Put $\sigma = \sigma(X)$ and $\tau = \tau(X)$.

By (2.3), we have $\sigma \leq \tau$.

By [**Trans. Ext.**] (1.11), we have a perfect surjection $X(\sigma) \to R^{\sigma}(X)$.

But $R^{\sigma}(X) = R^{\infty}(X) = \varnothing$, and so $X(\sigma) = \varnothing$ and hence $\tau \leq \sigma$.

Thus $\sigma(X) = \tau(X)$. ∎

References

[**Thesis**]   'Iterated Remainders in Compactifications' by P. P. Jackson

[**Trans. Ext.**]   'Transfinite Extensions' by P. P. Jackson, in this volume.

3. Forest Drainage

The Crychan Drainage Experiment After Felling

Hafren 4 Drainage Experiment

THE CRYCHAN DRAINAGE EXPERIMENT AFTER FELLING

Summary

The drainage characteristics of the Crychan experimental site after felling were examined by using a stochastic model for borehole water level hydrographs.

It was found that where cross drains were used a faster 'drainage rate' and a deeper 'base water table' were apparent.

Cultivation of the soil after felling was seen to impair the drainage characteristics specifically in terms of the 'drainage rate' and the 'computed infiltration at saturation'. This adverse effect was more pronounced on the plots without cross drains than on those with cross drains.

A comparison of pre-felling with post-felling drainage was made but the results were not clear.

The Crychan drainage experiment after felling

After the Crychan drainage experimental site had been felled eight extra boreholes were inserted into each subplot and half of each subplot was cultivated.

Daily borehole water level readings were taken over two 28-day periods together with daily rainfall readings.

The object of the experiment at this time was to see if the extra boreholes would improve precision in assessing the effect of cross drains on the drainage properties of the site; and to compare before felling with after felling drainage, and assess the effect of cultivation on drainage.

The experiment

The Crychan drainage experiment is a split-plot design experiment. Set out in four blocks with whole-plot factor drain depth at three levels: D0 = no drains, D2 = 2 ft drains, and D3 = 3 ft drains; the subplot factor is drain spacing at three levels, S1 = 50 ft spacing, S2 = 100 ft spacing, and S3 = 150 ft spacing.

In each subplot there are 12 boreholes in three rows of four. Situated one row towards the top, one in the middle, and one towards the bottom.

Each subplot has been split in half by cultivation or non-cultivation. This has been done randomly though not in strict accordance with a split-split plot randomisation.

Daily borehole water level readings and daily rainfall readings have been made for the two 28-day periods from 21st March to 17th April, 1983 and from 28th November to 25th December 1983.

Analysis of the experiment

The usual three parameter model for borehole hydrographs has been used with its estimated parameters λ, α, H and K used to characterise drainage properties. This is:

$Z_t = Z_{t-1} - \alpha X_t + (1 - \lambda)H + \text{error and } K = (1 - \lambda)H / \alpha.$

Examining the convergence of the parameter estimates for the first 28 day-period it was felt that the parameter estimates were still unstable after this period and so a further 28-day period of readings was obtained.

An ad hoc method of combining the two 28-day series into one 56-day series was used and the parameter estimates from this 56-day series used to characterise the drainage of the site. The parameter estimates from this 56-day series can be seen to have become stable after about 42 days readings.

After the removal of outliers - which are usually boreholes which show little or no response to rainfall and hence have a very large estimate of K - the Analysis of Variance tables for λ, α, H, and K. have been calculated, using means over sub-sub plots.

The effect of cross drains

The ANOVA table for λ shows a significant contrast between control and treated plots D0VD2D3 with p = .020. We see that λ is smaller on the plots with cross drains, that is mean D0 = 0.637, mean D2 = 0.559 and mean D3 = 0.596. This suggests a faster drainage on the treated plots. The 3ft ditches show no evidence of having any greater effect than the 2ft ditches.

The ANOVA table for α shows a significant D0VD2D3 contrast (p. = .001) and a significant interaction between the D0VD2D3 contrast and spacing. The values of α are greater where cross drains are used: mean D0 = 2.963, mean D2 = 4.282 and mean D3 = 4.372. However, on the treated plots α decreases with spacing.

The ANOVA table for H shows a highly significant D0VD2D3 contrast (p< .001). The means on the treated plots are greater than the untreated: mean D0 = 28.56, mean D2 = 37.57, and mean D3 = 40.42. There is no evidence that 3ft ditches have any greater effect than 2ft ditches.

The ANOVA table for K shows no significant depth, spacing or depth/spacing interaction effect.

The effect of cultivation

The Analysis of Variance tables for α and H show no significant cultivation effect.

The ANOVA for λ shows a highly significant cultivation effect (p< .001) with the cultivated plots having a higher value than the uncultivated that is mean C0 = 0.547 and mean C1 = 0.628. Thus drainage is slower on the cultivated plots.

The ANOVA for K shows a significant cultivation effect (p = .011) with K having larger values on the uncultivated plots. That is mean C0 = 4.35 and mean C1 = 3.74. Thus the computed infiltration at saturation is less in the cultivated plots and so the cultivated plots are less well drained. The tables of means show that K is reduced by cultivation more on the untreated plots than on the treated plots. That is mean D0C0 = 4.32 mean D0C1 = 3.13 whereas mean D2C0 = 4.40 mean D2C1 = 4.05 and mean D3C0 = 4.34 mean D3C1 = 4.03.

The effect of felling

Prior to felling there were four boreholes in each subplot making a total of 144 boreholes. Of these after cultivation a total of 72 were left in uncultivated sub-subplots. We use these 72 boreholes to compare the drainage before and after felling.

We use a paired comparison test. That is, for each borehole, we take the difference of the pre-felling value of the drainage parameters and the post-felling value. We then use a t-test to see if the mean of the resulting distribution is different from zero.

After allowing for outliers we obtain the following table:

	mean	variance	missing values	n
λ	0.239	0.021	15	57
α	2.552	20.845	10	62
H	14.904	190.156	0	72
K	-1.603	5.675	15	57

And also:

$t_\lambda = 0.239 / (\sqrt{(0.021 / 57))} = 12.4516$

$t_\alpha = 2.552 / (\sqrt{(20.845 / 62))} = 4.40125$

$t_H = 14.904 / (\sqrt{(190.156 / 72))} = 9.17095$

$t_K = -1.630 / (\sqrt{(5.675 / 57))} = -5.16586$

All of which are significant at the 1% level.

Thus in terms of our parameters before felling λ is larger, α is larger, H is larger, and K is smaller.

Discussion

In analysing the effect of cross drains we see that λ is smaller and H is larger when cross drains are used. Both of these indicate better drainage on the treated plots. Specifically, a faster 'drainage rate' and a deeper 'base water table'. However the effect on α seems difficult to interpret. Theoretically α is related to the porosity of the soil and the interception loss of rainfall, neither of which would be affected by the use of cross drains. It seems likely therefore that this effect is really a by-product of the estimation procedure whereby the value of α is smaller or larger according to some compensatory mechanism in estimating the parameters. Clearly this apparent treatment effect on α is the reason for $K = (1 - \lambda)H / \alpha$ showing no treatment effect where one might expect to find such an effect.

In assessing the effect of cultivation this difficulty has not arisen. Cultivation is seen to hinder drainage specifically in terms of the 'drainage rate' and the 'computed infiltration at saturation', and does not appear to have affected the 'base water table'.

The comparison between pre-felling and post-felling drainage requires caution. Theoretically the model used has parameters which do not depend upon the particular series of rainfall events under which the observations were taken. However, in practice, when different length series with different rainfall events are used it seems unlikely that identical parameter estimates would be obtained, even if there had been no change in the drainage characteristics of the site. With this in mind we see that the 'base water table', H, is deeper prior to felling - this could be attributed to the transpiration of the trees - however, the interpretation of the differences in the other parameters is not clear.

Conclusions

We have seen that plots which use cross drains show a faster 'drainage rate' and a deeper 'base water table' than untreated plots. Thus drainage has been improved by the use of cross drains. There is no evidence that 3ft drains provide any greater improvement than do 2ft drains.

We have seen that post-felling cultivation impairs the drainage in terms of the 'drainage rate' and the 'computed infiltration at saturation'. The untreated plots seem to be more adversely affected by post-felling cultivation than do those plots with cross drains.

HAFREN 4

DRAINAGE EXPERIMENT

Summary

We have analysed some recent data obtained from the drainage experiment Hafren 4. This data includes daily borehole water level readings, daily rainfall readings, measurement of the diameter at breast height of sample trees, and measurement of the height of some sample trees.

By modelling the boreholes hydrographs a quantitative characterisation of the drainage properties of the site is obtained. Statistical analysis shows that the use of cross ditches can lower the 'base water table' and increase the 'computed infiltration at saturation'.

Analysis of the growth data shows an increase in mean basal area and 'top' height when cross drains are used. There was no detectable difference between the effect of 2ft and 3ft deep cross drains.

Further analysis suggests that the increase in mean basal area is accounted for by the effect of the cross drains in lowering the 'base water table'; and the increase in 'top' height is additionally accounted for by the increase in the 'computed infiltration at saturation'.

The effect of fertiliser applied is significant and seems to have a relatively greater effect on the control plots (ie those without cross drains) than on the treated plots. However the yield from the treated (ie with cross drains) plots even without fertiliser was greater than from the untreated plots with fertiliser.

Introduction

The Hafren 4 drainage experiment is one of a series of experiments which were undertaken with the following objectives.

1.   To investigate the effect of two drain depths and three drain spacings on:-

    (a)   the soil moisture regime;

    (b)   the root development and stability;

    (c)   crop volume production.

2.   To assess which is the most economical and effective combination of drain depth and spacing.

3.   To assess the effect of fertiliser, in the presence and absence of drainage, on growth when applied to different ages of crop.

We have taken new measurements on this site relating to drainage and growth. We have quantified the drainage properties of the site by means of a stochastic model.

A statistical analysis of the growth data has been made and related to the drainage properties of the site.

The drainage experiment at Hafren

The experiment

The drainage experiment at Hafren is set in a P51 crop of Sitka spruce (Picea sitchensis (Bong) Carr.) in which treatment drains were put in in July 1966.

The experiment is of a split-split plot design set out in three blocks. The whole-plot factor is drain depth at three levels: D0=no drains (control), D2=2ft drains and D3=3ft drains. The whole-plots are separated by main drains, which run perpendicular to the contour. The treatment drains run parallel to the contour and divide the whole-plots into three subplots. The subplot factor is drain spacing at three levels: S1= 50ft spacing, S2=100ft spacing and S3=150ft spacing. The sizes of the subplots are 0.15 acre, 0.30 acre and 0.45 acre according as S1, S2 or S3 spacing is used.

The subplots are further divided into two (perpendicularly to the contour) according to the application of fertiliser; a sub-subplot factor at two levels: F0 no fertiliser

100

applied and Fl fertiliser applied. The fertiliser used was Fison's special 12:24:12 at 4cwt per acre in April 1967.

Borehole data

In each subplot, situated centrally between the treatment drains, a row of 4 boreholes has been inserted. Thus two boreholes lie in an unfertilised half and two boreholes lie in a fertilised half. There are 108 boreholes in total.

For the period of 42 days from 24.11.84 to 4.1.85 daily readings of the depth to water in these boreholes were taken using an electronic dipstick.

Also for the same period daily rainfall readings were taken as an average of two gauges.

Growth data

In the vicinity of each borehole 12 trees were selected and their diameters at breast height (diam bh) measured. The tree with the largest diam bh out of the 12 was then measured for height. These measurements were taken in May 1985.

Other data

For use as possible concomitant variables, measurements of the slopes of the sides of each subplot were taken. Also, measurements of the litter/peat depth at 12 points within each subplot were made.

Characterizing Drainage

Modelling the borehole data time series

The hydrologic time series of daily depth to water in a borehole will be used to try to quantitatively describe the drainage characteristics of the experimental units.

To do this a first order autoregressive model as described by Rennolls, Carnell and Tee 1980 is used to relate borehole level response to rainfall events. Specifically the three parameter model used is:

$$Z_t = \lambda Z_{t-1} - \alpha X_t + (1 - \lambda)H - V_t.$$

where Z_t is the depth to water in the borehole at the end of the t-th day, X_t is the amount of rain which fell on the t-th day, and V_t is an error term.

λ is an index of the rate at which the phreatic surface falls, the drainage rate. (a low valve means rapid drainage).

α   is an index of the drainable pore space, it also includes the interception loss.

H   is the asymptotic depth to which the phreatic surface apparently decays in the absence of rain.

The computed infiltration at saturation (c.i.s.) is the amount of rainfall in cm/day required to just maintain complete saturation of the soil under the model assumptions. Its value, K, is obtained from the formula:

$$K = (1 - \lambda)H / \alpha$$

The estimated values of λ, α, H and K, are used as a parametric characterisation of the drainage process.

Fitting the model

Parameter estimates $\lambda^\wedge$, $\alpha^\wedge$ and $H^\wedge$ are obtained by least squares estimation. The estimated expected trajectories can then be compared with the actual recorded water level time series. This can be done graphically for a selected number of boreholes. The resulting time series plots show no strong reason for rejecting the model. Therefore the model fit is taken as being adequate. (See figure 2).

Examination of the estimated parameters, $\lambda^\wedge$, $\alpha^\wedge$, $H^\wedge$, $K^\wedge$ shows that 3 boreholes out of the 108 have estimates which are clear outliers. These boreholes have been excluded from the subsequent analysis.

For a specified borehole the parameter estimates can be examined for 'convergence'. This is done by re-estimating the parameters when 6,7,8... ,42 days of the experiment only are considered. Graphical plots of these estimates show them becoming stable after about 30 days readings have been taken. (See figure 3).

Analysis of the experiment

1 Drainage characteristics

To see whether the use of treatment drains has changed the drainage characteristics of the experimental units we have calculated Analysis of Variance tables with the means of λ, α, H and K, taken over the experimental units, used as variates. It is thought that fertiliser will not affect the drainage characteristics so that here the subplots form the experimental units.

The orthogonal contrasts to compare control with treated and to compare 2ft drains with 3ft drains are included in the Analysis of Variance table i.e. DOVD2D3 (-2,1,1) and D2VD3 (0,-1,1).

The Analysis of Variance tables with λ or α as variate show no significant treatment effect, contrast, or interaction.

The Analysis of Variance table for H shows a very significant (p=.002) contrast between control and treated, D0VD2D3; however D3 shows no significant improvement over D2. There is a significant spacing effect (p=.038), and the contrast D0VD2D3 shows a significant interaction with space (p=.022). Examining the tables of means we see that mean H decreases with increased spacing over the treated plots, but no such trend is apparent over the control plots - where we would not expect to find such a trend. Also we can see that H is larger on the treated plots than on the control plots: mean D0=29.4, mean D2=55.4 and mean D3=57.3. See table 1.

The Analysis of Variance table for K shows a significant control v treated contrast, D0VD2D3, (p=.022). The contrast D2VD3 is not significant. A spacing effect is close to significance (p=.075) but there is no depth space interaction. The tables of means show that K is larger on the treated, D2,D3, plots than on the control, D0, plots: mean D0=5.15 Mean D2= 7.72, and mean D3 =8.80. See table 2.

Thus in terms of the 'base water table' and the 'computed infiltration at saturation' it seems that the treatments have effected the drainage. A lower base water table and a higher computed infiltration at saturation are apparent when treatment ditches are used. Also the smaller the spacing the deeper the base water table.

If the average slope of a subplot is used as a covariate then there is no significant covariate effect using λ, α, or H. However a significant covariate effect (p=.034) is found in the block-depth stratum for K. This use of a covariate brings the D2VD3 contrast to near significance at p=.068.

2. Growth data

The diameter at breast height (diam bh) measurements have been converted to basal area. Averages of basal area over the experimental units have been taken that is an average of 24 trees over each experimental unit.

Examining the Analysis of Variance table for the variate basal area we see a highly significant control v treated contrast, D0VD2D3, (p=.002), a significant spacing effect (p=.037), a highly significant fertiliser effect (p<.001), and a significant interaction between the D0VD2D3 contrast and fertiliser (p=.031). See table 3.

From the tables of means we see that basal areas are greater on the treated plots than the controls: mean D0= .018, mean D2=.024, and mean D3= .023. The basal areas decrease with increased spacing, however this spacing effect may be a spurious as it seems to be concentrated on the control plots where no spacing effect would be expected. See table 4.

The interaction between the D0VD2D3 contrast and fertiliser shows that the application of fertiliser has had a relatively greater effect on control plots than on treated plots. However the mean basal areas for treated plots without fertiliser are still greater than the mean basal areas for control plots with fertiliser e.g. mean D0F1 = .021 mean D2F0 = .023. For the control plots, mean D0F0 = .016 mean D0F1 = .021, and for the treated plots mean D2F0 = .023, mean D2F1 = .026, mean D3F0 = .023 and mean D3F1 = .024.

Of the 12 trees measured for diam bh in the vicinity of each borehole, the tree with largest dbh was also measured for height. These 'top' height measurements have been averaged over the experimental units i.e. for each experimental unit a 'top' height' is the average of 2 trees height measurements.

The Analysis of Variance for height shows a significant control v treated, D0VD2D3 contrast (p=.029), and a significant interaction between the D0VD2D3 contrast and fertiliser, (p=.046). See table 5.

The treated plots have a greater average 'top' height than the controls: mean D0 = 15.71, mean D2 = 17.68 and mean D3 = 17.87. The fertiliser has apparently a greater effect on the control plots than on the treated plots, the means being larger when fertiliser is applied: mean D0F0 = 15.15 and mean D0F1 = 16.26 compared with mean D2F0 = 17.51 and mean D2F1 = 17.86, and mean D3F0 = 18.08 and mean D3F1 = 17.66. See table 6

Average litter/peat depth was tried as a covariate, but for both basal area and height no significant covariate effect was found. Although with height as variate a nearly significant effect (p = .074) was found.

Average slope of the subplots was also tried as a covariate. For basal area there was no significant covariate effect. Although height showed a significant covariate effect (p = .017) in the blocks.depth stratum. This brought the D0VD2D3 contrast to a higher significance level (p = .002),

To see, whether the effect of the treatments on basal area and height, is accounted for by the effect of the treatments on the drainage characteristics. We have produced analysis of covariance tables of basal area and height when the drainage parameter, or the log of the drainage parameters are used as covariates.

For basal area we see that using H as a covariate will increase the p value of the control v treated, D0VD2D3, contrast from p = .002 to p = .295, and log H will increase the p value to p = .328. Thus the greater basal area on the treated plots is accounted for by the effect of the treatment on the drainage parameter H.

However the p value for the spacing effect p = .049 is only changed to p = .087 with H as covariate or p = .069 with log H as covariate. So the spacing effect is not accounted for by the treatment effect on H, but we have already seen that the spacing effect for basal area may be spurious anyway.

When height is examined by including drainage parameters as covariates we see that the p value for the control v treated, D0VD2D3, contrast, p = .029 is increased by including H to p = .635 and increased by using K to .693. Thus the greater 'top' height on the treated plots, D2 and D3, is accounted for by the effect of the treatments on the drainage parameter H and K.

Conclusions

We have seen that it is possible to quantify the drainage characteristics of a site by means of an autoregressive model for borehole readings time series.

Analysis of the borehole readings showed that the treated, D2 or D3, plots have a lower base water table, H, and a higher computed infiltration at saturation, K, than is apparent on the control, D0, plots. There was no evidence that the 3ft, D3, ditches had any greater effect than the 2ft, D2, ditches.

The analysis of basal area (average of 24 trees per experimental unit) shows that the treated, D2, D3, plots have produced a significantly greater yield (b.a.) than the control, D0, plots. Also the effect of the fertiliser applied is highly significant, with fertiliser having a relatively greater effect on the control, D0, plots. However yields (b.a.) for the treated plots, D2, D3, are higher than the controls, D0, even when fertiliser is applied. The increase in yield (b.a.) in the treated, D2, D3, plots over the control, D0, plots is accounted for by the effect of the treatment ditches in lowering the base water table.

Similarly, the analysis of 'top' height (average of 2 trees per experimental unit) shows the treated D2, D3 plots have produced significantly greater 'top' height than the control, D0, plots. Also the fertiliser has a relatively greater effect on the control, D0, plots than on the treated, D2, D3, plots. The increase in 'top' height in the treated, D2, D3 plots over the control, D0, plots is accounted for by the effect of the treatment ditches in lowering the base water table and increasing the computed infiltration at saturation.

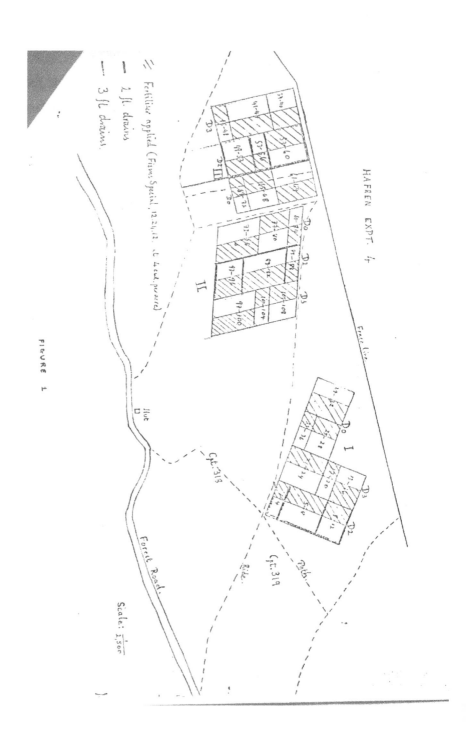

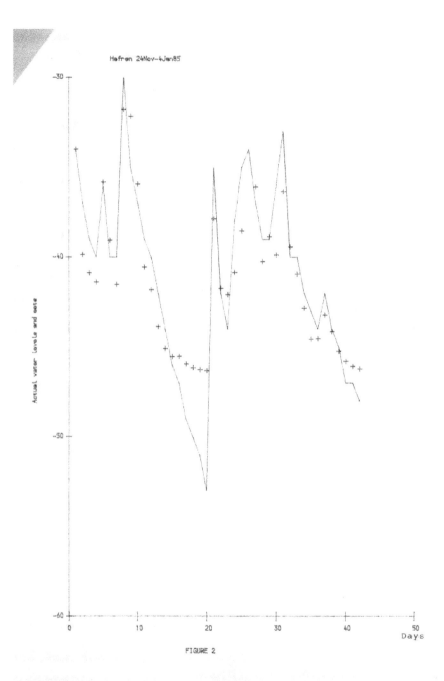

FIGURE 2

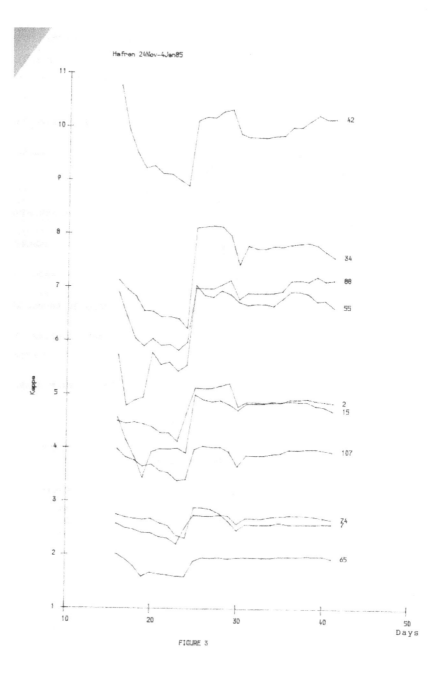

Hafren 24Nov–4Jan85

FIGURE 3

TABLE 1

****** ANALYSIS OF VARIANCE ******						
VARIATE H						
SOURCE OF VARIATION	DF	SS	SS%	MS	VR	F PR
BLOCKS STRATUM	2	99.61	1.73	49.80		
BLOCKS.DEPTH STRATUM						
DEPTH	2	4353.87	75.49	2176.94	29.072	0.004
D0VD2D3	1	4337.44	75.21	4337.44	57.925	0.002
D2VD3	1	16.44	0.29	16.44	0.220	0.664
RESIDUAL	4	299.52	5.19	74.88		
TOTAL	6	4653.39	80.68	775.57		
BLOCKS.DEPTH.SPACE STRATUM						
SPACE	2	270.89	4.70	135.44	4.367	0.038
DEPTH.SPACE	4	371.36	6.44	92.84	2.993	0.063
D0VD2D3.DEV	2	332.39	5.76	166.19	5.358	0.022
D2VD3.DEV	2	38.98	0.68	19.49	0.628	0.550
RESIDUAL	12	372.22	6.45	31.02		
TOTAL	18	1014.47	17.59	56.36		
GRAND TOTAL	26	5767.47	100.00			
GRAND MEAN				47.4		
TOTAL NUMBER OF OBSERVATIONS				27		

****** TABLES OF MEANS ******

VARIATE H

GRAND MEAN	47.4		

DEPTH	D0	D2	D3
	29.4	55.4	57.3

SPACE	S1	S2	S3
	51.4	43.7	47.0

SPACE DEPTH	S1	S2	S3
D0	31.3	21.1	35.9
D2	59.3	53.2	53.6
D3	63.7	56.7	51.4

TABLE 2

****** ANALYSIS OF VARIANCE ******						
VARIATE KAPPA						
SOURCE OF VARIATION	DF	SS	SS%	MS	VR	F PR
BLOCKS STRATUM	2	17.055	7.38	8.527		
BLOCKS.DEPTH STRATUM						
DEPTH	2	63.189	27.33	31.595	7.177	0.047
D0VD2D3	1	57.965	25.07	57.965	13.167	0.022
D2VD3	1	5.224	2.26	5.224	1.187	0.337
RESIDUAL	4	17.609	7.62	4.402		
TOTAL	6	80.799	34.95	13.466		
BLOCKS.DEPTH.SPACE STRATUM						
SPACE	2	45.284	19.59	22.642	3.242	0.075
DEPTH.SPACE	4	4.253	1.84	1.063	0.152	0.953
D0VD2D3.DEV	2	0.474	0.21	0.237	0.034	0.967
D2VD3.DEV	2	3.778	1.63	1.889	0.270	0.768
RESIDUAL	12	83.817	36.25	6.985		
TOTAL	18	133.354	57.68	7.409		
GRAND TOTAL	26	231.207	100.00			
GRAND MEAN				7.23		
TOTAL NUMBER OF OBSERVATIONS				27		

******* TABLES OF MEANS ******			
VARIATE KAPPA			
GRAND MEAN	7.23		
DEPTH	D0	D2	D3
	5.15	7.72	8.80
SPACE	S1	S2	S3
	8.99	5.92	6.77
SPACE	S1	S2	S3
DEPTH			
D0	6.66	3.93	4.88
D2	9.09	6.96	7.12
D3	11.22	6.86	8.32

TABLE 3

****** ANALYSIS OF VARIANCE ******						
VARIATE BASAL AREA (SQUARE CM) (MEAN PER TREE)						
SOURCE OF VARIATION	DF	SS	SS%	MS	VR	F PR
BLOCKS STRATUM	2	8448.7	9.65	4224.3		
BLOCKS.DEPTH STRATUM						
DEPTH	2	35775.6	40.88	17887.8	25.382	0.005
D0VD2D3	1	34715.2	39.67	34715.2	49.259	0.002
D2VD3	1	1060.4	1.21	1060.4	1.505	0.287
RESIDUAL	4	2819.0	3.22	704.7		
TOTAL	6	38594.6	44.10	6432.4		
BLOCKS.DEPTH.SPACE STRATUM						
SPACE	2	6994.8	7.99	3497.4	4.404	0.037
DEPTH.SPACE	4	3684.1	4.21	921.0	1.160	0.376
D0VD2D3.DEV	2	582.0	0.66	291.0	0.366	0.701
D2VD3.DEV	2	3102.1	3.54	1551.1	1.953	0.184
RESIDUAL	12	9530.8	10.89	794.2		
TOTAL	18	20209.7	23.09	1122.8		
BLOCKS.DEPTH.SPACE.FERT STRATUM						
FERT	1	9410.2	10.75	9410.2	30.655	<0.001
DEPTH.FERT	2	2307.7	2.64	1153.9	3.759	0.043
D0VD2D3.DEV	1	1683.1	1.92	1683.1	5.483	0.031

D2VD3.DEV	1	624.6	0.71	624.6	2.035	0.171
SPACE.FERT	2	1939.1	2.22	969.6	3.159	0.067
DEPTH.SPACE.FERT	4	1078.5	1.23	269.6	0.878	0.496
D0VD2D3.DEV.DEV	2	384.8	0.44	192.4	0.627	0.546
D2VD3.DEV.DEV	2	693.7	0.79	346.9	1.130	0.345
RESIDUAL	18	5525.5	6.31	307.0		
TOTAL	27	20261.1	23.15	750.4		
GRAND TOTAL	53	87514.0	100.00			
GRAND MEAN				220.7		
TOTAL NUMBER OF OBSERVATIONS				54		

TABLE 4

******* TABLES OF MEANS ******			
VARIATE BASAL AREA (SQUARE CM) (MEAN PER TREE)			
GRAND MEAN	220.7		
DEPTH	D0	D2	D3
	184.8	244.0	233.2
SPACE	S1	S2	S3
	236.0	217.3	208.7
FERT	F0	F1	
	207.5	233.9	
SPACE	S1	S2	S3
DEPTH			
D0	205.2	175.3	173.9
D2	269.7	239.6	222.9
D3	233.1	237.1	229.4
FERT	F0	F1	
DEPTH			
D0	163.7	205.9	

120

D2		230.6		257.5	
D3		228.1		238.3	

FERT		F0		F1	
SPACE					
S1		220.8		251.2	
S2		198.0		236.7	
S3		203.7		213.8	

SPACE	S1		S2		S3	
FERT	F0	F1	F0	F1	F0	F1
DEPTH						
D0	183.0	227.5	152.2	198.4	156.0	191.9
D2	250.8	288.6	214.7	264.4	226.3	219.5
D3	228.5	237.6	227.1	247.2	228.7	230.0

TABLE 5

****** ANALYSIS OF VARIANCE ******							
VARIATE 'TOP' HEIGHT (METRES)							
SOURCE OF VARIATION	DF	SS	SS%	MS	VR	F PR	
BLOCKS STRATUM	2	50.9034		26.49	25.4517		
BLOCKS.DEPTH STRATUM							
DEPTH	2	51.8503	26.99	25.9252	5.573	0.070	
D0VD2D3	1	51.5292	26.82	51.5292	11.078	0.029	
D2VD3	1	0.3211	0.17	0.3211	0.069	0.806	
RESIDUAL	4	18.6063	9.68	4.6516			
TOTAL	6	70.4567	36.67	11.7428			
BLOCKS.DEPTH.SPACE STRATUM							
SPACE	2	2.8929	1.51	1.4464	0.543	0.595	
DEPTH.SPACE	4	10.2394	5.33	2.5598	0.960	0.464	
D0VD2D3.DEV	2	0.5959	0.31	0.2979	0.112	0.895	
D2VD3.DEV	2	9.6435	5.02	4.8217	1.809	0.206	
RESIDUAL	12	31.9836	16.65	2.6653			
TOTAL	18	45.1159	23.48	2.5064			
BLOCKS.DEPTH.SPACE.FERT STRATUM							
FERT	1	1.6017	0.83	1.6017	1.860	0.189	
DEPTH.FERT	2	5.2900	2.75	2.6450	3.072	0.071	
D0VD2D3.DEV	1	3.9675	2.06	3.9675	4.607	0.046	

D2VD3.DEV	1	1.3225	0.69	1.3225	1.536	0.231
SPACE.FERT	2	0.4169	0.22	0.2085	0.242	0.787
DEPTH.SPACE.FERT	4	2.8564	1.49	0.7141	0.829	0.524
D0VD2D3.DEV.DEV	2	1.8385	0.96	0.9192	1.067	0.365
D2VD3.DEV.DEV	2	1.0179	0.53	0.5090	0.591	0.564
RESIDUAL	18	15.5000	8.07	0.8611		
TOTAL	27	25.6650	13.36	0.9506		
GRAND TOTAL	53	192.1410	100.00			
GRAND MEAN				17.09		
TOTAL NUMBER OF OBSERVATIONS				54		

TABLE 6

******* TABLES OF MEANS ******			
VARIATE 'TOP' HEIGHT (METRES)			
GRAND MEAN	17.09		
DEPTH	D0	D2	D3
	15.71	17.68	17.87
SPACE	S1	S2	S3
	17.41	16.87	16.98
FERT	F0	F1	
	16.91	17.26	
SPACE	S1	S2	S3
DEPTH			
D0	16.21	15.49	15.42
D2	18.63	17.23	17.19
D3	17.38	17.90	18.33
FERT	F0	F1	
DEPTH			
D0	15.15	16.26	

D2	17.51	17.86
D3	18.08	17.66
FERT	F0	F1
SPACE		
S1	17.28	17.54
S2	16.58	17.17
S3	16.89	17.07

SPACE	S1		S2		S3	
FERT	F0	F1	F0	F1	F0	F1
DEPTH						
D0	15.95	16.47	14.92	16.07	14.58	16.25
D2	18.27	19.00	17.12	17.33	17.15	17.23
D3	17.62	17.15	17.70	18.10	18.93	17.73

4. Database Development

Use of Rushmore Technology in Browse Edit Windows

A Custom Built General Reporting Facility

USE OF RUSHMORE TECHNOLOGY IN BROWSE EDIT

WINDOWS

A large research organization required a system to keep track of components sent off site on loan to other companies or organizations for various reasons such as repair or warranty. For each item sent there is an officer in the organization responsible for periodically assessing if the item has been returned, is still on loan or needs hastening.

A small database system was required, for the use of by an inventory officer, who would periodically issue reminders, extracted from the database, to the various responsible officers. These would then return them with their comments; and the inventory officer would update the data tables with review dates, comments, etcetera, and a flag to indicate if the item had been returned. We decided to implement this using FoxPro 2.6 for DOS.

On the Administration network we had access to a data table of staff members keyed on a six character (digit) staff number field. It was decided to use this as the source for the names and details of the Responsible Officers, as this contained their room numbers, which could be used to address the reminders. We created a loan item table keyed on a dispatch note number, of seven characters, from the dispatch notes used when an item was shipped. Then to keep track of the review dates we created a reviews table keyed on the seven character dispatch note number concatenated with the review date as a date string (i.e. using the dtos() function).

Thus our small database consisted of the following tables:

NAME.DBF

STAFFNO	Character 6
SURNAME	Character 20
INITIALS	Character 8
TITLE	Character 4

...

LOANITEM.DBF

DESPATCH	Character 7
CONTRACT	Character 8
STAFFNO	Character 6
RETURNED	Character 1

...

REVIEWS.DBF

DESPATCH	Character 7
REV_DATE	Date
REMARKS	Memo

...

We implemented the system and pretty soon had a loan item data table containing several thousand records, loaded from a selection of dispatched items from an existing hierarchical database system. However, as the returns came in from the responsible officers, the inventory officer needed to find by dispatch number then edit the returned flag for each item and we were asked to find a way to ease the tedium and time spent on this process.

Our solution to this was to use a Browse For window which would display the items for a responsible officer, in dispatch note number order, and permit editing of the returned flag only. However initially the delay in displaying only those records for a particular responsible officer was unacceptably long. To overcome this we created a tag on the staffno field in the loanitem data table which enabled Rushmore Optimisation to come in to play and the display became instantaneous. One point we did notice, however, was that use of the 'Exactly Like' operator (i.e. = =) in the Browse For clause seemed to disable the Optimiser so it was necessary to use the 'Like' operator (i.e. =).

We give the (annotated) coding for the routine below:

```
******************************************************************
procedure dereturn

minput = .f.
is_there = .f.
```

```
select loanitem

set order to tag namedesp   && order by staffno and despatch number

itrec = recno()


m->surname = space(20)  && Prepare to select responsible officer

activate window inp_win

clear

@ 2, 2 say "Enter Surname " get m->surname picture "!!!!!!!!!!!!!!!!!!!!!"

read

deactivate window inp_win


do pickname  && Routine to enable user to select an officer from the NAME table

if abort  && User cancelled in pickname routine

   return

endif


select loanitem

set order to tag namedesp

seek m->staffno + space(7)   && m->staffno set up in pickname routine

if eof()

  go bottom

endif


if loanitem->staffno <> m->staffno  && Officer chosen has no loanitems

  abort = .t.
```

130

```
    do errrep with "No Items for " + trim(m->surname) + " " + trim(m->initials)
else
  abort = .f.
endif


if .not. abort  && We need to allow the user to do the editing
  select name
  set order to tag namekey  && Ordered by staffno field
  select loanitem
  set relation to staffno into name


  set confirm on


  browse fields returned:H='Return':P='!':V=returned $ "YN ",;
    despatch:H='Number':R,contract:H='Contract':R,;
    name->surname:H='Surname':R,name->initials:H='Init':R,;
    ...
    noappend nodelete for staffno = m->staffno ;
    title 'Return Loan Items : <CTRL><END> - exit' ;
    freeze returned ;
    preference "returned" window brws_win


  set relation to
  set confirm off
```

```
endif


if abort  && Need to find loanitem for display

 select loanitem

 set order to tag despatch

 go itrec

endif


do deveri  && Redisplay screen


return

*****************************************************************
```


This completes our listing of the coding for the routine. The use of the freeze

clause in the Browse command allows the user to do his or her editing without having

to move from the desired column. The inventory officer was more than satisfied with

our solution as it reduced his workload considerably.

A CUSTOM BUILT GENERAL REPORTING FACILITY

This facility was devised and programmed in response to the requirements of users in a large Personnel Department, but the technique employed can and has been used on other databases both simple and complex. The programming was done in FoxPro 2.6 for DOS. The users required a facility to produce reports from their database according to various and variable selection criteria and having only those fields they selected displayed and in the order in which they chose. They had had some training in the use of FoxPro but found the basic facilities provided to be too difficult, cumbersome, time consuming, and technical for their requirements and, indeed, not capable of doing some of the tasks they required.

Our solution to this was to provide a reporting facility, customised to the structure of the their database and easy to use. This has been found to cater for a whole host of their needs; and eased the need for many specially programmed static reports, which would otherwise have had to be made. Of course, static reports are still used and are therefore required from time to time.

We begin with two data tables which we copy to the users local drive or a personal network drive and these copies are used only for the duration of one run of the report program. The first table, which we shall call GENREP.DBF, is an empty table, which contains as fields all the fields or calculated fields, which may be required to be printed out on the report. The field names need not correspond to those in the database as GENREP.DBF will be filled with the required data by our program and scanned when the report is output. It also contains a field ,sortkey, and tag (of size

Character 100) which our program dynamically fills so that the table is in the required order. The second table, which we shall call GENFIELD.DBF is not empty and has the following structure:

GENFIELD.DBF

CHOICE	Character	2		
SORTBY	Character	2		
FIELD	Character	10		
DESC	Character	30		
LEN	Numeric	3	dec.	0
HEADING	Character	15		
TYPE	Character	1		
LIKE	Character	10		
EQUAL	Character	10		
ABOVE	Character	10		
BELOW	Character	10		
IN_LIST	Character	40		
NOTINLIST	Character	1		

For each field in the GENREP.DBF data table, there corresponds a record in the GENFIELD.DBF data table where Genfield.Field contains the field name as in the structure of GENREP.DBF, Genfield.Desc contains a description of the field contents, Genfield.Heading contains the heading for the field (to be put on the report), Genfield.Len contains the maximum of the field width and the heading length, and

Genfield.Type contains "C" if the field in GENREP.DBF is Character, "D" if Date, "N" if Numeric, "M" if Memo etcetera. The other fields in GENFIELD.DBF are initially blank.

The user is provided with a means of browsing the GENFIELD.DBF table and adding entries to the fields which as yet are still blank. By putting the numbers 1,2,3, ... in the Genfield.Choice field the user is enabled to select the fields he or she requires to be output and the order in which they are to appear on the report. Similarly entries in the Genfield.Sortby field enable the user to select the ordering of the data which is to appear in the report. Entries in the other fields are fairly self-explanatory but we shall describe the principles just for completeness. Thus entries in Genfield.Like restrict the report to those values of the corresponding field which are 'Like' the entry (i.e. as if the = operator was used). Similarly for Genfield.Equal but there the 'Exactly Like', (i.e.= =), operator is used. Genfield.Above and Genfield.Below are used to restrict entries to those which are greater than or equal to (i.e. >=), or less than or equal to (i.e. <=), the user input value respectively. The Genfield.In_list field may contain a comma separated list of values, and will then restrict to field values taking one of the values in the list (by use of the inlist() function). By putting an entry in the Genfield.Notinlist field we can reverse the logic and restrict to field values which do not correpond to those given in the Genfield.In_list entries. A useful trick implemented with this is that if the Genfield.Notinlist field contains an entry, but the Genfield.In_list field is blank then only non-blank entries for the corresponding field will be selected.

These user options will be stored in arrays by our program, which will then use macro substitution to build up the code required to implement the choices selected by the user. Details of this method will be shown below in the abbreviated extract of the code.

The user is also given the facility to make selections from criteria peculiar to the structure of the database and take the form of 'often required' selection criteria. For example, in our application, staff who are or were with the organization on a given date. These type of selections have to be specially programmed into the routine which extracts the data and fills the (copy of) GENREP.DBF.

We now give a skeleton coding for the routine:

```
*****************************************************************
procedure genmain

* Open all relevant data tables and set up relations

...

* Create copies of GENREP.DBF and GENFIELD.DBF on the user's personal drive.

...

* Allow the user to select the required fields to be output and the required ordering of
* the data by filling in the appropriate Genfield.Choice and Genfield.Sortby fields as
* described above.

...

* Obtain the application specific selection criteria from the user.
```

…

* Allow the user to enter values into the appropriate Genfield.Like,

* Genfield.Equal, Genfield.Above, Genfield.Below,

* Genfield.In_list, and Genfield.Notinlist fields, to complete his/her

* selection of data to extract.

…

* Declare arrays aChoice, aHeading, aTypes, aLens, aSortby, aSorttypes

* (of size 20 in our application)

* Declare array aSels (of size 100 in our application)

…

* Set up these arrays with values according to the users selection.

select Genfield

set order to tag choice && Ordered by val(choice)

go top

nChoice = 0 && Counter for number of output fields selected.

if seek(1) && by pass all blank choices

  do while .not. eof()

    nChoice = nChoice + 1

    aChoice(nChoice) = Genfield.Field

    aLens(nChoice) = Genfield.Len

    aHeading(nChoice) = Genfield.Heading

    aTypes(nChoice) = Genfield.Type

```
    skip

  enddo

else

* Code to cope with case user selected no fields.

…

endif


select Genfield

set order to tag sortby  && Ordered by val(sortby)

go top


nSortby = 0  &&  Counter for number of fields in ordering key.

if seek(1)  && By pass blank selections.

  do while .not. eof()

    nSortby = nSortby + 1

    aSortby(nSortby) = Genfield.Field

    aSorttypes(nSortby) = Genfield.Type

    skip

  enddo

else

* Code to cope with case user selected no ordering

…

endif                          `


* Now build up cKeyfield variable ready for macro substitution
```

```
cKeyfield = " "

for ii=1 to nSortby

  do case

    case aSorttypes(ii) = = "C"

      cKeyfield = cKeyfield + "+ m->" + aSortby(ii)

    case aSorttypes(ii) = = "D"

      cKeyfield = cKeyfield + "+ dtos(m->" + trim(aSortby(ii)) + ")"

    case aSorttypes(ii) == "N"

      cKeyfield = cKeyfield + "+ str(m->" + trim(aSortby(ii)) + ",7,2)"

  endcase

endfor


* Now set up aSels array ready for macro substitution


select Genfield

set order to 0  && In record number order

go top


nSels = 0  && Counter for number of entries in aSels array


scan  && Go through all records in the Genfield table


  if Genfield.Like <> " "

    nSels = nSels + 1
```

```
   do case

     case Genfield.Type = "C"

       aSels(nSels) = "m->" + trim(Genfield.Field) + "=" + ;

                       "" + trim(Genfield.Like) + ""

     case Genfield.Type = "D"

       aSels(nSels) = "m->" + trim(Genfield.Field) + "=" + ;

                       inscurly(trimGenfield.Like))   && inscurly() adds { and } if needed

     case Genfield.Type = "N"

       aSels(nSels) = "m->" + trim(Genfield.Field) + "=val(" + ;

                       Genfield.Like + ")"

   endcase

 endif


 if Genfield.Equal <> " "

   nSels = nSels + 1

   do case

*  Similar case statement to above but with the "=" operator replaced by

*  the "= =" operator.

...

   endcase

 endif


 if Genfield.Above <> " "

*  Similar code to above but using ">=" operator

...
```

```
    endif


if Genfield.Below <> " "
*  Similar code to above but using "<=" operator

...

    endif


if Genfield.In_list <> " "
  nSels = nSels + 1
  do case
    case Genfield.Type = "C"
      aSels(nSels) = "inlist(m->" + trim(Genfield>Field) + "," + ;
          insquo(Genfield.In_list) + ")"  && insquo() adds quotes into the list
    case Genfield.Type = "D"
      aSels(nSels) = "inlist(m->" + trim(Genfield.Field) + "," + ;
          inscurly(trim(Genfield.In_list)) + ")"  && inscurly() adds brackets into list
    case Genfield.Type = "N"
      aSels(nSels) = "inlist(m->" + trim(Genfield.Field) + "," + ;
                    Genfield.In_list + ")"
  endcase


if Genfield.Notinlist <> " "
  aSels(nSels) = ".not." + aSels(nSels)
endif
endif
```

* Now add code to deal with 'not blank' choice as explained above

if Genfield.In_list = " " .and. Genfield.Notinlist <> " "

  nSels = nSels + 1

  do case

    case Genfield.Type = "C"

    aSels(nSels) = "m->" + trim(Genfield.Field) + "<> ' '"

    case Genfield.Type = "D"

    aSels(nSels) = "m->" + trim(Genfield.Field) + "<> { / / }"

    case Genfield.Type = "N"

    aSels(nSels) = "m->" + trim(Genfield.Field) + "<> 0"

  endcase

 endif

endscan && Scan of Genfield data table

* Select main data table to scan through

…

scan

* Code to scan through the main database setting memory variables to the

* values required for each entry in the GENREP.DBF data table, and implementing

* the specialised selection criteria for the database as chosen by the user.

…

142

* Now put in the aSels array of selection criteria using macro substitution

```
lCondloop = .f.
for ii=1 to nSels
  if .not. ( &aSels(ii) )
    lCondloop = .t.
  endif
endfor


if lCondloop
  loop
endif


* Now set the ordering of the data according to the users choice


m->sortkey = &cKeyfield


select genrep
append blank
gather memvar memo


* Now reselect main data table
...
endscan   && Scan of main data table
```

* We now have all the values gathered into GENREP.DBF with the data selected

* as required by the user

* Call the routine to produce the printed output

do gensub

* Close all data tables and delete temporary copies of GENREP and GENFIELD

...

RETURN

****

procedure gensub

* To scan the now filled GENREP.DBF table and produce the printed report

* Code to set printer to the required filename set up page length etcetera

...

printjob

* Code to set printer to required orientation and font size

...

* Code to call on page routine at required page length

...

```
select genrep

set order to tag sortkey

go top

nTotal = 0  && Counter for number of records reported on


scan


  scatter memvar memo

  nTotal = nTotal + 1


  ?

  nCol = 1

  for ii=1 to nChoice  && Print out user selected fields only

    do case

      case aTypes(ii) = "C"   && Character field

        ?? m->&aChoice(ii) at nCol

      case aTypes(ii) = "D"   && Date field

        ?? dtoc(m->&aChoice(ii)) at nCol

      case aTypes(ii) = "N"   && Numeric field

        ?? strzbl(m->&aChoice(ii),aLens(ii),2) at nCol   && strzbl() is str() with 0 blank

      case aTypes(ii) = "M"   && Memo field

        nLin = memlines(Genrep->&aChoice(ii))

        if nLin >= 1

          ?? mline(Genrep->&aChoice(ii),1) at nCol

          for jj=2 to nLin
```

```
        ? mline(Genrep->&aChoice(ii),jj) at nCol

    endfor

  endcase

  nCol = nCol + aLens(ii) + 2   && Leave 2 spaces between fields on report
  endfor


endscan  && Scan of GENREP.DBF


* (Note: This coding tacitly assumes only one memo field selected by the user

*   and that this is the last field in the user list of choices).


on page

endprintjob


* Code to send printer reset to file and set printer off

…

RETURN

******************************************************************

procedure genpage


* This is the 'on page' routine


eject page

private kk, nPcols
```

* Code to output the heading for the report as required by the organisation

…

?

nPcol = 1

for kk=1 to nChoice

  ?? trim(aHeading(kk)) at nPcol

  nPcol = nPcol + aLens(kk) + 2

endfor

?

nPcol = 1

for kk=1 to nChoice

  ?? replicate("-",aLens(kk)) at nPcol

  nPcol = nPcol + aLens(kk) + 2

endfor

?

…

RETURN

****

function insquo

* To insert quotes into a comma delimited string of values

* ie JOHN, PAUL,GEORGE becomes "JOHN","PAUL","GEORGE"

…

RETURN retvar

```
*********************************************************************f
```

function inscurly

* To insert curly brakets in a comma delimited string of dates

* i.e. 31/01/95,22/12/95,01/01/96 becomes {31/01/95},{22/12/95},{01/01/96}

...

RETURN retvar

```
*********************************************************************
```

function strzbl

* To imitate the FoxPro str() function but to return a blank string if the passed

* value is zero

...

RETURN retvar

```
*********************************************************************
```


This completes our abbreviated listing of the code. The program could be enhanced by, for example, making an output file suitable for loading into an Excel spreadsheet. Also the number of decimals required for numeric fields could be added if required. (In our application this was always 2 and so not taken into account.)